Thinking Big

Thinking Big

How the Evolution of Social Life Shaped the Human Mind

Clive Gamble
John Gowlett
and Robin Dunbar

57 illustrations

First published in 2014 in hardcover in the United States of America by
Thames & Hudson Inc., 500 Fifth Avenue, New York, New York 10110

thamesandhudsonusa.com

Library of Congress Catalog Card Number 2013950868

ISBN 978-0-500-05180-1

Printed and bound in India by Replika Press Pvt. Ltd.

Contents

Preface

Thinking big is a very human thing to do. We have remarkable imaginations that take us back into the past and out into the future. We feast on films and literature and ruminate on the creativity of the human mind. We take innovations in social networking in our stride as we adapt to living in mega-cities. We work in global economies and access what we know of these big worlds from 24/7 news. But, for all this big thinking, we remain in some essentials small-minded. We have a cognition that equips each of us to deal best with small numbers of other people. While the size of our populations has grown exponentially, we remain, at core, the product of the small social worlds of our evolutionary history.

What we set out to explore in this book is the story of how our big-thinking social brain evolved. We do this by looking at ourselves and our closest living relatives – the apes and monkeys – as well as at the skulls and artifacts of our fossil ancestors. There is a link between the two and that is the size of the brain and the size of the small social communities they live in. We will examine the idea that it was our social lives that drove the growth of our most distinctive feature: the human brain.

Our ability to think big is part of our evolutionary story. It was with the desire to know more about this vital ingredient of being human that we undertook a seven-year project (2003–2010) funded by the British Academy as a celebration of its centenary. We called the project *Lucy to Language: The Archaeology of the Social Brain* and in the pages that follow you will learn much about Lucy's journey from an ancestor with a small brain to a global species with an open mind.

The Lucy project was collaborative and interdisciplinary. Its rationale was underpinned by Sir Adam Roberts, President of the British Academy, when he wrote about the public value of the humanities and social sciences as follows: 'The humanities explore what it means to be human: the words, ideas, narratives and the art and artifacts that help us make sense of our lives and the world we live in; how we have created it, and are created by it. The social sciences seek to explore, through observation and reflection, the processes that govern the behaviour of individuals and groups. Together, they help us to understand ourselves, our society and our place in the world.' This sums up the intent of our Lucy project – to explore the past and the present to provide a fuller account of where we have come from and why we act the way we do.

Our greatest thanks go to the British Academy and their decision to fund a project that drew together the humanities and the social sciences. We were also extremely fortunate in having a Steering Committee of Garry Runciman, Wendy James and Ken Emond, who read all our reports, attended all our conferences and through their enthusiastic support and advice added immeasurably to the success of the research undertaken. We would also like to thank David Phillipson FBA for all his help in furthering our research in Africa.

We benefited throughout from the encouragement and wise advice of the project's Honorary Fellows: Leslie Aiello, Holly Arrow, Filippo Aureli, Larry Barham, Alan Barnard, Robin Crompton, William Davies, Bob Layton, Yvonne Marshall, John McNabb, Jessica Pearson, Susanne Shultz, Anthony Sinclair, James Steele, Mark van Vugt, Anna Wallette, Victoria Winton and Sonia Zakrzewski.

One of our goals was to build the next generation of researchers in human evolution to work across the humanities and social sciences. We were very fortunate in having such a talented group of postdocs and postgrads, many of whom are now embedded in the universities of the world. Our postdoctoral research fellows were Quentin Atkinson, Max Burton, Margaret Clegg, Fiona Coward, Oliver Curry, Matt Grove, Jane Hallos, Mandy Korstjens, Julia Lehmann, Stephen Lycett, Anna Machin, Sam Roberts and Natalie Uomini; our research assistants were Anna Frangou and Peter Morgan; and our postgraduate students were Katherine Andrews, Isabel Behncke, Caroline Bettridge, Peter Bond, Vicky Brant, Lisa Cashmore, James Cole, Richard Davies, Hannah Fluck,

Babis Garefalakis, Iris Glaesslein, Charlie Hardy, Wendy Iredale, Minna Lyons, Marc Mehu, Dora Moutsiou, Emma Nelson, Adam Newton, Kit Opie, Ellie Pearce, Phil Purslow, Yvan Russell and Andy Shuttleworth. In relation to African research and fire studies John Gowlett also thanks especially Stephen Rucina, Isaya Onjala, Sally Hoare, Andy Herries, James Brink, Maura Butler, Laura Basell, National Museums of Kenya and NCST Kenya; also Nick Debenham, Richard Preece, David Bridgland, Simon Lewis, Simon Parfitt, Jack Harris, Richard Wrangham and Naama Goren-Inbar.

Funding for the fellowships and studentships came principally from the British Academy Centenary Project, and our meetings, fieldwork and study leave were made possible by grants administered under the British Academy's Research Professorship, Small Grants, Conference and Exchange funding programmes. We were also very grateful for additional funding from the Arts and Humanities Research Council, Economic and Social Research Council, Engineering and Physical Sciences Research Council, Natural Environment Research Council, the Leverhulme Trust, the Boise Fund and the EU-FP6 and FP7 programmes. We also received generous support from our host institutions, Oxford University, Liverpool University, Royal Holloway and Southampton University.

Any long-running project is marked by the natural rhythms of life. Five babies were born and none of them are called Lucy! We are happy to say their social brains are developing nicely.

Clive Gamble
John Gowlett
Robin Dunbar

1

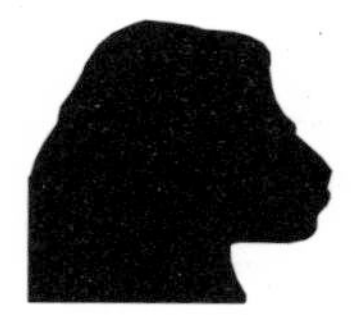

Psychology meets archaeology

The history of human evolution is an iconic story that never ceases to mesmerize and enthral. Buried in our past is one of the triumphs of evolution, the story of how a common-or-garden African ape began to change both its body form and the way it lived its life – and how in doing so, it eventually became the dominant species on earth. It is only within the last century that we have really come to appreciate the grandeur of this story and the moments of uncertainty and near-extinction that threatened it.

From small beginnings

Some 7 million years separate us from the time that the ancestors of humans and chimpanzees were a single species: a small, undistinguished African Miocene ape.* We finished that part of our story in the last 5000 years as the only animal to have settled all the terrestrial habitats of the earth, from the tropical forest to the arctic tundra and from high mountain plateaux to small islands in the remotest oceans. During that long history the size of our brains trebled and our technology progressed from simple stone tools to digital marvels. We walked upright, spoke, made art in profusion and crafted worlds of enormous imaginative complexity in the name of religion, politics and society. Truly, we are no longer apes.

* The Miocene dates from about 23 to 5.3 million years ago. It is followed by the Pliocene (5.3 to 2.6 million years ago) and the Pleistocene (2.6 million to 11,700 years ago).

For most of these 7 million years we were not alone. Where our remote ancestors lived they often shared the space with other closely related species. This ancient pattern began to change within the last 100,000 years when people like us, modern humans, moved out of Africa and through the Old World. Older species like the Neanderthals of Europe and Western Asia were displaced and became extinct. These same modern people also passed beyond the boundaries of the Old World, peopling for the first time Australia and the Americas. By the time the last Ice Age ended 11,000 years ago, we were the only species in town; *Homo sapiens* was now alone, in an evolutionary sense.

Soon we also became a global species. The move to farming led in one direction to cities, civilizations and a massive increase in population. And in another direction the domestication of plants provisioned the voyages into the remote Pacific, beginning 5000 years ago, while harnessing the power of animals made it possible to traverse cold and hot deserts. No wonder then that the European voyages of discovery found people everywhere; what is more, these explorers tested time-and-again the historical circumstance of *Homo sapiens* as a single biological species through successful, if not always consensual, interbreeding.

We still carry this 7-million-year history in our bodies and our brains. The scientific insights that arise by comparing our own anatomy and that of the great apes have been essential for understanding the process of evolution, and a revolution in genetics has opened up new evidence for tracing ancestral lineages using both modern and ancient DNA. Fossil skeletons, skulls and teeth have also received forensic attention for the evolutionary information they contain. At the same time, archaeologists have charted the development of technology and tackled key issues concerning diet and the behaviour that ensured a reliable food supply. The result is a much richer and better-understood record of our earliest history.

We began our scientific careers in the late 1960s when the landscape of human evolution was very different. There were few fossils and science-based techniques of dating (led by radiocarbon) were still in their infancy. Getting to see sites and materials was both difficult and expensive until the jumbo jet transformed international travel in 1970. Computers filled entire basements and had to be programmed with punch cards. There were no touch screens or search engines and as postgraduate students the greatest luxury we had was a photocopier, expensively producing images on shiny paper.

It is easy to be dazzled by the rate of technological change and the speed with which new data about our earliest origins have built up. The beginning always seems small by comparison with the present. But small should not be taken to mean unimportant. We will show in this book that for all their sophistication those material changes are still directed at solving some age-old issues of being human. These concern our social lives, which we believe have been largely ignored in the study of our origins.

Our major proposition in this book is that a link has always existed between our brains, or more precisely the size of our brains, and the size of our basic social units. We see this link as essential to understanding our evolution as a single, global species that can live in cities the size of Rio de Janeiro, drawing daily on vast amounts of information to manage our lives. But inside today's global citizen is a social being who carries forward a social life that in its basics is very similar to one 5000 or 50,000 years ago. At the core of this social life is the observation that a limit of about 150 exists in terms of the size of your social network. This is known as Dunbar's number, as one of us, Robin Dunbar, did the research that established the figure. This limit is almost three times greater than the chimpanzee's, which immediately raises the evolutionary question of how did this increase occur? It also begs another question: if the limit is 150, then how come we can live in such large cities and align ourselves to massively populous nations the size of China or the United States?

Our aim in this book is to trace the evolutionary journey from our small beginnings to the present position. Our principal guides are psychologists and archaeologists, although many other disciplines have been involved. With our social perspective on human evolution, we have set out to learn about the following central issues:

- Is there a limit in our brains, our cognitive ability, that restricts the size of the social groups we can live in?
- If so, how did our cognitive ability evolve to cope with ever greater numbers of people, as societies grew from the small social worlds of hunters to today's mega-cities?

- Given that our ancestors had much smaller brains than ours, what do we mean when we talk of a social life in the remote past?
- Will it ever be possible to say when hominin brains became human minds?

The list above could of course be much longer, but these core questions indicate our interest, first and foremost, in the social rather than in a history of technology or the architectural details of fossil skulls. They also point to our concerns with cognitive matters, the business of understanding how and why we think and act the way we do. Our approach is underpinned by evolutionary theory and our goal is to apply the insights from an experimental subject such as psychology to a historical discipline such as archaeology. This is rarely attempted and never easy. But first some background.

The germ of an idea takes shape

In 2002, the British Academy, the UK's national body for the humanities and social sciences, launched a competition for a research project to celebrate the centenary of its foundation. It proposed to give the largest single grant it had ever made to a flagship project in the humanities and the social sciences. Although our individual perspectives and interests had been quite different, the three of us had spent most of our professional lives immersed in the story of human evolution. One of us was a Palaeolithic archaeologist with a primary interest in Africa, one a social archaeologist with a special interest in late Palaeolithic societies in Europe, the third an evolutionary psychologist with a principal interest in human and primate behaviour.

It seemed to us, contemplating the opportunities that such a project might offer, that we were perfectly placed to rise to the challenge that the British Academy had thrown down. We had the single biggest question one could ever ask (how did we come to be human?) and we could bring novel expertise to bear on the question. Where past studies of human evolution had been obliged to concentrate on the limited physical evidence that was available (the stones and the bones), we were fortuitously in a position to exploit recent findings about social behaviour and brain evolution in a way that might illuminate the significance and meaning of the stones and the bones. Moreover, archaeology was in the humanities half of the Academy and psychology in the social sciences half, so we could bridge the divide over which the Academy presided, offering an iconic example of how to do interdisciplinary research. We quickly became galvanized with enthusiasm, put our heads together and sent in a bid.

The possibilities that such an endeavour offered seemed positively limitless. The academic world was just beginning to grapple with the integration of psychology and archaeology. The previous decade had witnessed the creation of cognitive archaeology under the driving force of the British archaeologist Colin Renfrew and the American archaeologist Thomas Wynn. The main focus of this approach had been understanding the cognitive demands of toolmaking and the production of works of art. But we felt that recent developments in our understanding of the behaviour of our nearest living cousins, the monkeys and apes, and in the processes underpinning important areas such as brain evolution, would enable us to go one step beyond to say something about the social life of hominins (see Table 1.1), and to do so much further back in time than most cognitive archaeologists had previously dared to go. In particular, the theory that had become known as the social brain hypothesis – the idea that the brain had evolved to allow animals like monkeys and apes to handle an unusually complex social world – offered novel insights and rich seams to exploit in the exploration of hominin social evolution.

Anthropoids	All primates (monkeys and great apes and their fossil ancestors), hominins and humans
Hominids	All great apes (gorillas, orang utans, chimpanzees, bonobos, gibbons), hominins and humans
Hominins	All our fossil ancestors *(Ardipithecus, Australopithecus, Homo)*
Humans	Only modern humans, *Homo sapiens*
Anatomically modern humans	*Homo sapiens* but without substantial evidence for our cultural accoutrements (art, burials, ornament, musical instruments)

Table 1.1: *Common terms in human evolution.*

Our bid was grandly entitled *Lucy to Language: The Archaeology of the Social Brain*. Lucy was the iconic early australopithecine fossil that had been unearthed by the palaeoanthropologist Don Johanson and his team in the deserts of northeastern Ethiopia in 1974 (it was named after the Beatles' song *Lucy in the Sky with Diamonds*, which had been playing on a tape recorder when the fossil was unearthed). Lucy and her family had lived around 3.5 million years ago, and marked the earliest well-documented hominins. Since the australopithecines still shared many similarities with our common ape ancestors – at least in terms of brain capacity – it seemed like the obvious place to start our story. Language marked the appearance of modern humans, our own kind, and seemed like a natural end point. And so the project acquired its name.

After submitting our proposal, we could only sit back and wait. It is not easy to get funding for research in the sciences or the humanities these days in any country, so we were under no illusions about the outcome. The funding rates of the UK research councils are notoriously low, with only about 10 per cent of proposals actually receiving grants – despite the fact that almost all of those submitted involve exciting, novel and innovative science. We fully expected to be presiding over yet another failed bid. So it was with some surprise and excitement that we heard that our project had been shortlisted for the final interview stage. We were in with a chance!

In the end, of course, this particular story had a happy ending, or we wouldn't now be writing this book. Ours was the project that was selected by the British Academy as its Centenary Research Project. It turned out that the competition had been much tougher than we had imagined. There had been more than 80 other proposals submitted. Many other potentially exciting projects had faced disappointment, with all the attendant wailing and gnashing of teeth as is inevitable under such circumstances. But, with money for a seven-year project assured, all we had to do was put together a team of exciting young researchers and venture purposefully into the unknown. *Thinking Big* is the story of our project.

The social brain and its evolution

The centrepiece of our project was the social brain hypothesis. This had taken its first hesitant steps in the 1970s when it was pointed out that monkeys and apes had much bigger brains relative to body size than any other animals. Pondering this, a number of primatologists more or less independently suggested that this was probably because monkeys and apes live in unusually complex societies. Later, during the 1980s, the primatologists Andy Whiten and Dick Byrne, of the University of St Andrews, suggested that what made primate societies so complex was the behaviour of the animals themselves. A monkey group wasn't like a beehive, which has enormous structural complexity arising from the fact that different individuals are programmed to perform different tasks. Beehives are largely the outcome of strict chemical management of behaviour: individual bees do not choose to adopt the roles of worker or drone or queen, but rather are obliged to behave this way by a combination of their genes and the chemical signals imposed on them by the rest of the hive. Monkeys, by contrast, are individuals and, within the constraints of their individual psychologies, adjust their behaviour according to the exigencies of the particular circumstances in which they happen to find themselves.

The complexity of primate societies is created by the subtleties of how the individuals interact with each other. And, as every field primatologist will tell you, it is the soap opera of daily life in a monkey group that creates both its fascination and its intricacy. Whiten and Byrne lit on the fact that monkeys and apes are forever deceiving and outwitting each other in a perennial attempt to steal a fast one in the great race of life. A monkey might surreptitiously hide a desirable fruit to prevent another seeing it; or it might give an alarm call to distract everyone else from noticing that it had found a particularly nice bulb that needed some time to dig out of the ground. Whiten and Byrne named this the Machiavellian intelligence hypothesis, in honour of Niccolò Machiavelli, the Italian Renaissance political philosopher whose iconic book *The Prince* had spelled out the devious political strategizing that would best guarantee a late medieval ruler success and long life.

Because some people objected to the implied suggestion that primate politics were driven by the same deviousness as human politics, the name for the theory was later changed and the social brain hypothesis was born. In part, this was in recognition that it wasn't just the behavioural complexity of monkeys and apes that was at stake, but also the sizes of their groups. The seal was imposed on this story during the 1990s, when it was shown that the average size of a species' social groups correlated with the size of its brain (see Figure 1.1); or to be more precise, correlated with the size of its neocortex (literally 'new cortex'), the outer layer of the brain, surrounding the so-called old brain (the brain stem and mid-brain, including the limbic system and the parts that run most of the body's autonomic activities). It is the neocortex that has exploded in size during the course of primate evolution. It was this massive expansion of the neocortex that was responsible for the fact that primates had larger brains than other mammals. The neocortex appears for the first time in the mammal lineage – although there is a comparable part of the brain in birds, too.

In the 60–70 million years of evolutionary time since the primates first appeared as a distinct group of mammals, the primate neocortex has gradually increased in size as species have evolved from one into another. It overlies what we might think of as the reptilian brain and it is what allows mammals to adjust their behaviour in more sophisticated ways to the exigencies of day-to-day circumstances. Although the complexity of behaviour and the psychology that underpins this is the key to the social brain story, the bottom line is that a species' brain size seems to impose a constraint on the size of its social groups. When groups exceed their species-typical limit, they begin to fall apart because the animals cannot manage to maintain coherent relationships with each other.

Two things seem to be important in this respect. One is the psychological sophistication of monkeys and apes, and their apparent ability to strategize and deceive. The other is the fact that this kind of social cognition is very expensive in computational terms: the neurons of the brain have to work hard. We will examine both in more detail in later chapters, but for the moment let it suffice to say that these two components are intimately related. We have been able to show that the skills on which the kind of sociality that humans have depend on a capacity known as mind-reading or mentalizing – the ability to understand or infer what another individual is thinking. This allows us to keep several people's intentions in mind at the same time, and so adjust our behaviour in such a way as to allow for their interests as well as ours when we act in a particular way. We have been able to show, in addition, that this capacity for coping with many individuals' mental states depends crucially on the volume of neural matter in particular parts of the neocortex. These regions, in the

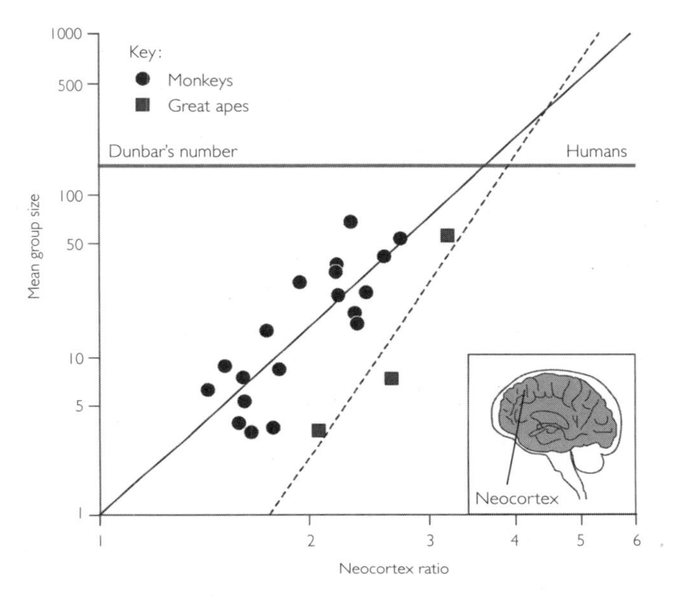

Figure 1.1: *Social group size for different monkey and ape species plotted against the species' relative neocortex size. The neocortex is the outer layer of the brain that is responsible for complex thought. The index of relative neocortex size (neocortex ratio) is neocortex volume divided by the volume of the rest of the brain: this allows us to standardize for differences in brain size.*

frontal and temporal lobes, form a network of neural clusters that are known to be crucial to mentalizing.

A key aspect of the social brain hypothesis of particular interest to our story is the fact that it makes a very specific prediction about the size of human groups. The equation that relates social group size in apes to a species' neocortex volume predicts that modern humans have a natural group size of about 150 individuals, the value that is, as we have seen, now known as Dunbar's number. One reason this is important for our story is the fact that the relationship shown in Figure 1.1 spans the sequence from chimpanzees (representative of the last common ancestor between great apes and humans) to modern humans: all the now-extinct hominin ancestors must lie along the line between these two points. Our task will be to figure out just where they fall and what the implications of this are for their social and mental lives.

Dunbar's number in the modern world

The social brain hypothesis predicts that humans have a natural grouping size of around 150. But is this really true? We only have to look around where most of us now live to see what is surely obvious: humans live in towns and cities that are considerably larger than 150 people. Indeed, many of the great modern cities of the world today number their citizens in the tens of millions. So how is it that the social brain equation gives us such a low number? Perhaps the theory is just wrong. Or perhaps what the theory is telling us is that the number of people that can be crammed together in a tangled mess of electricity wires, winding lanes and sewage pipes doesn't bear much relationship to the world of our social relationships. We can live in cities of tens of millions, but our personal social worlds – the worlds that consist of the people we actually know – are formed of a pint-sized 150 people. If this second suggestion is true, perhaps what Dunbar's number is all about is the limit on the number of individuals with whom we can have relationships. After all, if we think about what is involved in the original equation for monkeys and apes, it is the number of individuals who live together in a group on a daily basis. These groups are small, and, with a few exceptions, the animals see each other every day and all day. By no stretch of the imagination can everyone who lives in London, New York, Mumbai or Beijing possibly see each other every day, or even every month, or every year – and never mind meet the people from any of these other cities. And even if by some chance they did, they certainly wouldn't be able to remember who they all are. In actual fact, it seems that the limit on the number of faces we can put names to is around 1500–2000, and that is well below the size of even a small village in the modern world.

This puzzle set us thinking about the kinds of evidence we really needed to test the prediction of the social brain equation. There seemed to be two obvious places to look. One was in the kinds of small-scale societies in which we have spent most of our evolutionary history as a species. There are still quite a few of these around, but they are confined to the more obscure tribal societies on the margins of the modern world. They are the societies we find among hunter-gatherers, people like the Kalahari San of southern Africa, the Hadza of East Africa, or many of the rainforest tribal societies in South America, and, at least historically, among the Aboriginal peoples of Australia. The other possibility was to look at ourselves and our own personal social worlds, the network of individuals with whom we had personal social relationships.

The literature on community size in hunter-gatherer societies is slight, partly because anthropologists haven't been especially assiduous about collecting such data. There is, moreover, another source of confusion, and this is the fact that it isn't at first clear just what counts as a community among hunter-gatherers. Not unreasonably, many people have supposed that the basic group for hunter-gatherers is the set of people that camp together on a daily basis. This has a typical size of around 35–50, only a third of Dunbar's number. However, hunter-gatherer societies, like our own, consist of a variety of types of communities, which are typically organized as a hierarchy – families clustered within kinship groups, kinship groups clustered within villages, and villages clustered within larger regional groups. It is this last that turns out to be particularly interesting because it is this level of social organization, and it alone, that has a typical group size of the right sort of magnitude. The average is almost exactly 150. So we have some evidence that natural human community sizes are in fact of just the size predicted by the social brain hypothesis.

Figure 1.2 shows what you see if you look down on a population from above, the picture created by the way the people are distributed in geographical space. But what about the size of personal social networks, the social world seen from below, from the individual's point of view? Our first attempt to look at this involved Christmas cards. Each year, many of us sit down and spend a lot of time, effort and money sending cards

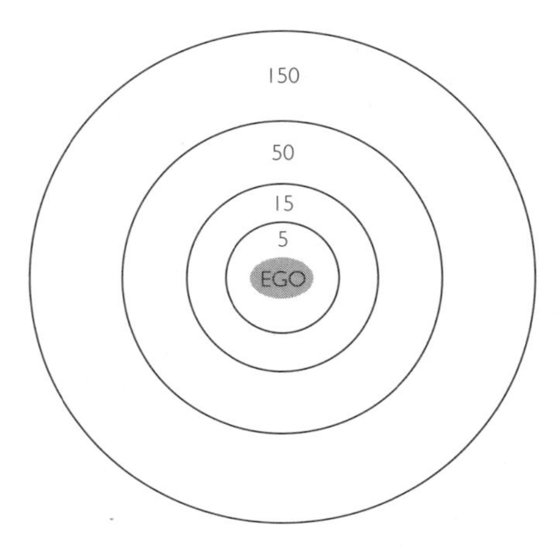

Figure 1.2: *The circles of friendship. The average person's social network consists of about 150 friends and family, arranged in a series of layers that correspond to different qualities of relationship, each of a very distinctive size. Each layer in this series is roughly three times larger than the layer inside it. The layers are roughly equivalent to intimate friends, best friends, good friends and friends.*

to people we want to keep in touch with. So one year, we asked about 45 people to keep a list of all the people in the households to which they were sending cards. Figure 1.3 shows the result. There was a fair bit of variation, but the average was in fact 154, about as close to the predicted value as one could possibly wish.

This prompted us to be a bit more ambitious, and over the following years we put together a large database involving 250 individuals who made complete lists of all the people whom they regarded as important in their personal lives. This was, it must be said, an arduous undertaking because we also asked them to tell us a lot of details about the individuals they listed – how they were related to them, when they had last seen them, how close they felt to them emotionally. But the end result was very rewarding because, again, the number 150 emerged as critical.

So between these various sources of data, we seemed to have strong evidence for the claim that our social world is quite small and limited to about 150 other people. From our big dataset, we were able to draw two more key conclusions. First, people varied quite a lot in the number of friends they had. In fact, the range of variation around the 150 mark is probably something like 100–200. Second, and this

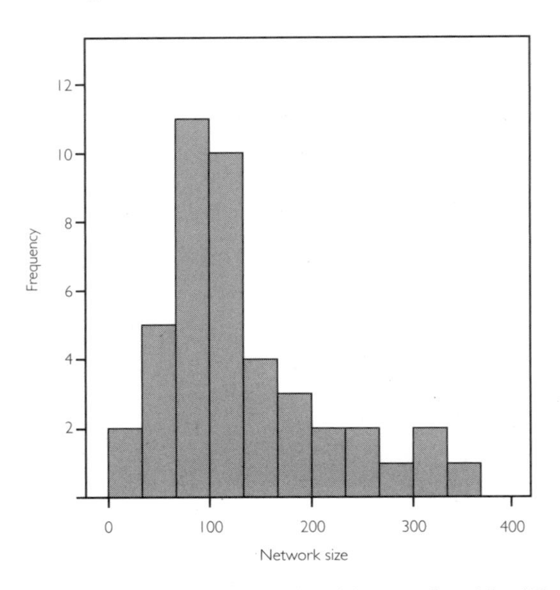

Figure 1.3: *The Christmas card lists of 45 people. While our social networks have a typical size of about 150 people, we vary a great deal in the number of friends and relations we have: some of us have very small networks (though we typically invest more time and effort in each relationship) and some have considerably larger ones (but typically invest less in each relationship).*

surprised us, about half the individuals included in people's 150s were family and half were friends.

Since all the people in our sample were Europeans (from the UK and from Belgium), we had assumed that family members would largely be represented just by close kin – mum, dad, brothers, sisters, grandparents, perhaps the odd aunt or uncle or cousin. But we thought the numbers would be quite small. Of course, kinship – and extended kinship – is important in traditional societies. Indeed, kinship has been the bread and butter of anthropological study since the discipline first got going a century-and-a-half ago. But we assumed that it was only traditional societies that still 'did' kinship in the extended sense: we in the developed world had abandoned such notions, preferring the advantages of social mobility over the ties of hearth and home, and as a result, while of course we still valued immediate family, our social worlds were dominated by friendships and acquaintances from work. This turned out not to be so. About half the people we include in our social network are members of our extended family. In fact, we could even show that people who come from big extended families list fewer friends in their social networks. So it seems that the figure of 150 is a real limit on the number of relationships you can have. You have just 150 slots, you give priority to family members first, and then fill the rest up with friends.

Of course, there are many ways to outflank the limits. You don't *have* to include your family in the list. Some people fall out with their close family and never see them again. Rather, the point is that, typically, people prioritize family above friends. If you don't have many family members, or you have fallen out with them, then you fill your slots with friends – or your favourite soap opera characters, or your pet, even your favourite potted plant if you genuinely feel you have a relationship with it! And of course, you can even include people who don't physically exist, like God or the saints, if you feel so disposed. The important thing to remember is that relationships, ties, bonds – call them what you will – are built up by us as we pursue our social lives. Who your mother and father are is given to you, as are those other biologically based kinship categories. But most of what we do is better described as a process of negotiation, building up and selecting a personal network of friends, loved ones and acquaintances. That number of 150 is a sample drawn from many possible choices.

The reason for this limit, Dunbar's number, will be explored in later chapters. But at this point we need to briefly introduce the idea of cognitive load, which is a way of thinking about our mental capacity to remember and act on information, in this case about others in our social community. We all know the feeling before an exam or an important presentation that our brains are full up, bursting with data, and as the number of our social relationships increases, so we are similarly faced with an overload issue. Can we remember names, personal histories and fulfil our obligations to others? It seems that the figure of 150 stretches to the limit our cognitive ability to remember, recall and react in consistent and socially productive ways. Cognitive load thus acts as a brake on our social ambitions.

The age of the past

Enter, at this point, archaeology – and the past. So far, we have given one side of the coin, the relationship between the brain and the social world in living species. Now we must turn to something just as crucial in our frame – the exploration of deep-history.* This is the forte of archaeology, which has its roots in the antiquarian movement that began more than 300 years ago and became one of the foci of intellectual curiosity throughout the Enlightenment. The archaeology we recognize today, however, was a product of the 19th century. During the first half of that century materials were classified into a three-age system – Stone, Bronze and Iron – which later would form the evidence for a simple evolution of society from hunters to farmers and ending with civilizations.

The question that archaeologists and geologists most keenly wanted an answer to was the antiquity of humans. Did they date to the Ice Age, which would make their origins very old, or only to the most recent geological period, as advocated by many who looked no further than the Bible for a chronology? The answer came 150 years ago, in 1859. Two Englishmen, Joseph Prestwich and John Evans, who went on to dominate their respective fields of geology and archaeology, were following up Frenchman Boucher de Perthes' claims that, in the Somme Valley in northern France, there was evidence that humans and extinct animals such as woolly rhino and mammoth had lived at the same time. On an April afternoon Prestwich and Evans found what they were looking for

* We use deep-history in preference to prehistory to describe the remote history of our earliest ancestors.

in a gravel pit at St Acheul in the suburbs of Amiens (hence the term 'Acheulean', subsequently adopted for this toolmaking epoch). They even took a photograph of the moment of discovery, which shows a stone tool, in situ, sticking out of the gravels in which they had also found the bones of extinct animals. Their results were instantly accepted back in London by the Royal Society and the Society of Antiquaries. The science of human origins had scored a notable success although intriguingly the stone handaxe that proved their point was lost from sight until Clive Gamble and Robert Kruszynski re-found it exactly 150 years later – it

Figure 1.4: *The St Acheul gravel pit in the suburbs of Amiens in the Somme Valley, France. This photograph, taken on 27 April 1859, shows a quarryman pointing to a handaxe (Figure 1.5) found in place in the Ice Age gravels.*

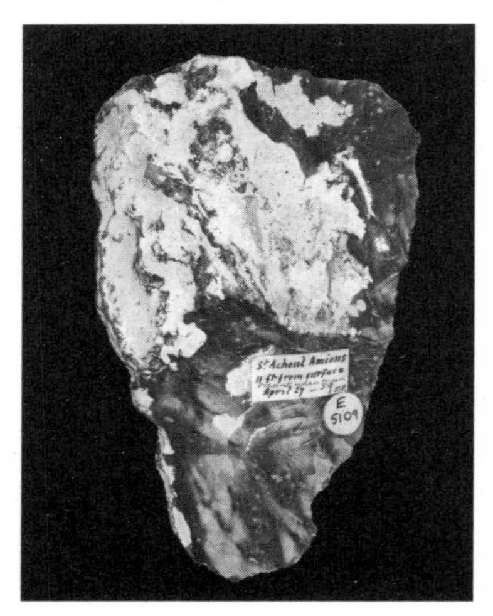

Figure 1.5: *A handaxe crucial to the history of Palaeolithic studies was recently relocated by Clive Gamble and Robert Kruszynski, still bearing its label of 1859.*

was safely stored in the collection of artifacts that Prestwich's widow had donated to what is now the Natural History Museum after he died in 1896. This was very definitely a stone that changed the world, shattering biblical chronologies and opening up a deep-history whose enormity, in the absence of dating techniques, could only be guessed at.

This same science of human origins did have an interest in ancient society. Sir John Lubbock subtitled his popular *Pre-Historic Times* of 1865 *as Illustrated by Ancient Remains and the Manners and Customs of Modern Savages*. People who lived by hunting and gathering, such as the Aboriginal Tasmanians, were seen as the modern representatives of the people who made the St Acheul handaxe – the people of Lubbock's Palaeolithic (Old Stone Age). They were distinct from the polished axe users of his later Neolithic (New Stone Age), when farming had replaced hunting as the means of subsistence. Such comparisons continued for many years until it was recognized that drawing direct parallels between the past and the present was both poor history and entirely misleading. Besides, such an approach erroneously assumed that people living today had not changed but were instead living fossils.

Archaeologists concentrated their subsequent efforts on amassing information, first from Europe and then from Asia and Africa. During the 20th century they became less interested in the social lives of these early humans and more in what they made and what they ate. But the social is inevitably at the heart of archaeology's ideas. It was dragged fully back into the picture by a brilliant Australian scholar, Vere Gordon Childe, in his *Social Evolution* published in 1951. He argued that archaeology must play the same role for anthropology as palaeontology for zoology, although for Childe society, as we shall see in Chapter 3, really took off with farming. However imperfectly the traces were preserved, what archaeologists studied was societies. And so, when Grahame Clark and Stuart Piggott went on to write a grand outline of human progress in 1965, they entitled it *Prehistoric Societies*.

These were the aspirations, but the frame of evolution – the necessary backdrop to our 'becoming human' – always depended on the earliest and sparsest evidence. It took pioneering resolve and major discoveries to catapult us forward into a modern evolutionary age. The greatest landmark was the discovery by Louis and Mary Leakey in 1959–60, at Olduvai Gorge in East Africa, of early hominin fossils, together with stone tools, in a setting that could be dated to nearly 2 million years ago.

At a stroke, the record became three times longer than most had thought possible. A time-depth to human origins was opened up that would have amazed Prestwich and Evans, who guessed a few hundred thousand years at most. That field season at Olduvai was the moment when the scale of the human past was mapped out in modern terms, with scientific methods of dating, such as potassium–argon, that are crucial for giving substance to the findings. Yet again other sciences, psychology included, were knocking on the door. Louis Leakey's timing was impeccable – he managed to get his key findings out for a volume published to celebrate the centenary of Darwin's *Origin of Species*. There, Leslie White conceived of 'four stages of minding' and Irving Hallowell wrote about 'self, society and culture'. So why did we not leap forward into a full appreciation of the early mind?

Part of the brake came from other seemingly positive developments. One was a revolution that stirred up archaeology in the 1960s – a revolution that became known as the 'New Archaeology'. For us, it was a two-edged sword. Some of its greatest exponents, among them Lewis Binford, exposed the limits of archaeological evidence by showing how selective preservation could distort the record, and how easy it is to create 'modern myths' about the ways that human life could be recovered. When stone tools were found clustered with animal bones, even with human fossils, you could not simply assume that this was a 'campsite' or even a 'kill site'. Too many other natural factors could produce the same configuration.

The human club and WYSWTW

Two aspects were frustrating not just for archaeologists, but for scientists in other disciplines trying to interpret the record. First, in the

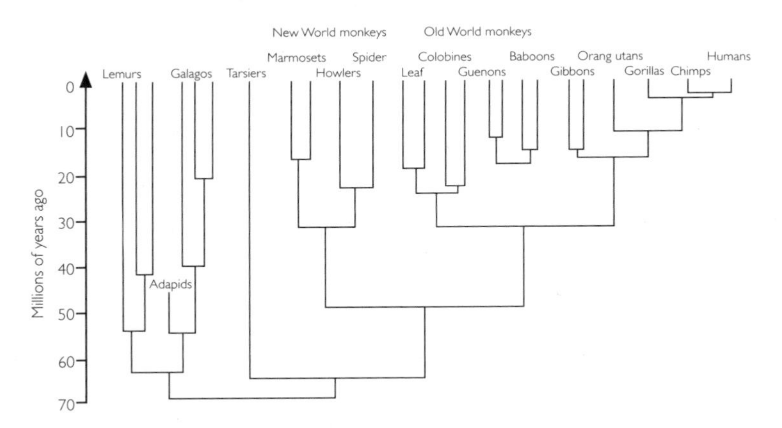

Figure 1.6: *A chart of primate evolution, showing the major divisions and the dates when they first appeared. On the left are the prosimians (represented today by lemurs and galagos); on the far right are the ape family (gibbons, orang utans, gorillas, chimpanzees and ourselves), with the New and Old World monkeys in between.*

New Archaeology the traditional narrative account was spurned on the grounds that our record was not history and could not make a connected story like history. Second, and more importantly, a view began to emerge insisting that *what you could not observe directly in the record you were forbidden even to think about*. At a stroke this idea, that What You See is What There Was (WYSWTW), excluded large areas of what it is to be human – emotions and intentions for example – from the scientific study of human origins. There were shades here reminiscent of the behaviourism that had dominated much of psychology from the First World War until the 1970s. Psychology is in origin about the working mind, and the behaviourists had argued that since the mind itself could not be observed directly, it was not even open for discussion. For human origins, too, many scholars felt that any advanced capabilities had to be demonstrated beyond reasonable doubt. Only if they expressed their ideas in material form – as art or skilled crafts – could earlier humans hope to be included in our modern club.

This idea of the modern human club has permeated archaeology since the 1970s, when a new phrase, 'the anatomically modern human', appears in the literature. This was coined to describe ancestors that looked like us and even had our genes but who did not behave like us. They had neither art nor basic architecture. Their burials were usually simple rather than rich, without grave goods or evidence for ceremony. The anatomically modern humans found in these burials date to between 200,000 and 50,000 years ago in Africa and the Middle East. With hindsight we can see that by membership of the modern human club what was actually meant was membership of the European club, where art and elaborate burials had been known for some time as a component of its Upper Palaeolithic.

But which sense of 'becoming human' do we mean here? For some people, there is only full membership: the ancestors have to be 'like us' to qualify, and that pretty much limits us to the last 200,000 years of hominin evolution. In this book, however, we are interested in a broader view, for we are primates, bound to our primate relatives such as the chimpanzee and the bonobo by a family tree that stretches back through several million years at least (see Figure 1.6). All of it is there to be explained, not merely the last few per cent.

Building the long record

Fortunately, that longer record has had a huge appeal. The new extended domains of deep-history made it seem as if archaeologists had pulled out a folding bed, and suddenly realized how much space they had. Pioneers possessed enormous enthusiasm and energy, launching fieldwork ventures that reached back 2 million years, and that drew in scientists from many other disciplines. Few people realize quite how much archaeology there is, and how many of its aspects interrelate with one another. Some biologists believe that archaeologists have just 'a few stones and bones'. But if we take the work of just some of the great Clarks of archaeology – Grahame Clark, the European prehistorian, J. Desmond Clark, the Africanist, F. Clark Howell, the palaeoanthropologist, or David L. Clarke, the brilliant theoretician who died at the tragically young age of 38 – you can see the immense variety of activity that has built up a record that is actually far too big for any one person to take in.

Figure 1.7: *The late Glynn Isaac was one of the leading thinkers in the generation of the 'New Archaeology'.*

One of the pioneers, Louis Leakey, played a crucial part in encouraging research into both past and present, including the great apes, and in this

Figure 1.8: *Olduvai Gorge in Tanzania was for many years the focus of research by the Leakeys. Streams have cut a great scar across the landscape exposing ancient lake beds and the activities of early hominins. The latter transported stones for toolmaking from the rocky hills in the background, providing one of the first clues that their networks of operation could be explored.*

way transformed our knowledge about the background of human evolution. Born in Kenya in 1903, he drew from his upbringing in Africa a particular appreciation of animal and human behaviour in wild environments. Alongside this bush experience he acquired a western education and was able to investigate sites of all periods, not just at the famous Olduvai Gorge, but also early Miocene ape sites around Lake Victoria, and later Stone Age sites in the Rift Valley – including, quite coincidentally, a site named Gamble's Cave!

While this very breadth, combined with a somewhat headstrong and maverick personality, often irritated his European colleagues, it allowed Leakey to focus on the essentials of actual life in the savannahs and forests as it might have been. His colleague, the anatomist Philip Tobias, once remarked that Leakey epitomized the idea that no one achieved much who never made mistakes, and commented especially on his vision. Leakey realized that we could only hope to understand extinct animals by using modern animals to help interpret 'their structure, functioning and behaviour'. He was far ahead of his time in appreciating how much the apes had to teach us, not just about themselves, but about the framework of human evolution and our own nature. Among his many activities, he paved the way for outstanding research by Jane Goodall on chimpanzees, Diane Fossey on gorillas and Birutė Galdikas on orang utans.

Through the 1960s to 1980s, there was a buzz of excitement in palaeoanthropology. In the field, it was a period of great international research expeditions, focused mainly on Africa, which drove our detailed knowledge of human origins back towards 4 million years, back towards the last common ancestor with the apes. Louis' son, Richard Leakey, and the energetic South African-born archaeologist Glynn Isaac opened up the expanses of East Turkana; F. Clark Howell and the Harvard group worked at Omo to the north; Don Johanson and Maurice Taieb led a breakaway expedition that culminated in the spectacular discoveries around Hadar in the north of Ethiopia – the home of Lucy and her *Australopithecus afarensis* kin. So rich were the findings in the same area by palaeoanthropologist Tim White and his team that many of the implications are still being unravelled. Nor was it just Africa: Europe was re-explored, and then the Far East and Australia. All have contributed crucial evidence that allows us to fashion the synopsis that follows.

The story of hominin evolution that has emerged from a century of fossil-hunting and careful fieldwork and museum analysis has, of course, changed considerably over the decades as new knowledge has become available. This is neither the time nor the place to recount that history of discovery. However, we do need at this point at least to sketch out the human story as we currently understand it, although parts of it will no doubt change as new fossils are discovered in the decades to come. But for present purposes we need a framework, summarized in Figure 1.6 and Table 1.2, round which to build the chapters that follow.

Our tale begins with the last common ancestor, or LCA, that our lineage had with the African great apes, and in particular with the chimpanzees (the ape to whom we are most closely related) some 7 million years ago. We have no idea what this ancestor looked like, because no fossils that can be identified as the LCA are known. It will not have looked exactly like a chimpanzee, since, like us, the chimpanzees themselves have had 7 million years of evolution of their own since the LCA roamed the forests of central Africa. Indeed, there is precious little to show for the first 2 million years post-LCA – a handful of recently discovered bones in East Africa and an impressive skull

1.	After at least 20 million years of apehood, our last common ancestor with the apes lived about 7 million years ago
2.	Upright walking and changes in the teeth started at least 4.4 million years ago, as the first hominins appear in the fossil record
3.	Stone technology became increasingly important from about 2.6 million years ago
4.	The brain began to enlarge significantly around 2.4 million years ago with the appearance of the earliest hominins who can be called *Homo*
5.	After 2 million years ago early humans moved out of Africa and around the Old World, reaching in places to above 55 degrees north
6.	Brain growth showed a marked rise with *Homo heidelbergensis* at 600,000 years ago; language was probably present but not necessarily as we know it
7.	Anatomically modern humans appeared in Africa about 200,000 years ago, as indicated by fossil and genetic evidence
8.	By 60,000 years ago (or earlier) modern humans had spread from Africa; they displaced existing hominins and moved into new lands such as Australia and, after 20,000 years ago, the Americas; the age of the single, global human species had begun
9.	Evidence for art, ornament and symbolic behaviour started in Africa with anatomically modern humans and before they left that continent; worldwide after 40,000 years ago, it increased in complexity and multiplied in frequency
10.	Major changes in the scale and organization of society began in the last 10,000 years when farming replaced hunting and gathering as the economic mainstay

Table 1.2: *Ten steps in the hominin–human story.*

from the edge of the Sahara desert in West Africa. What seems to mark these fossils out as different is an upright stance, a unique two-legged form of bipedal locomotion. All the other apes and monkeys walk on all-fours, and in the apes this takes the form of a distinctive body shape with short legs and long arms – a body shape associated with shinning up the vertical trunks of massive forest trees. Our lineage seems to have been distinguished right from its outset by an upright body with longer legs and shorter arms, a body shape adapted to striding across the open ground between the trees. While the earliest hominins don't have quite as elegant a bodyline as we do, they are nonetheless characterized by this distinctive trait. Indeed, it is just about the only trait that actually distinguishes the members of the hominin family from the other great apes.

However, broadly speaking, we seem still to be dealing with ecological apes. But they were apes that had diversified into a variety of niches that would not have been habitable for living apes, because fruits and soft shoots would not have been available in the extremes of dry and wet seasons. They include the robust australopithecines with their massive teeth, the so-called gracile australopithecines, with less heavy jaws and teeth, and the more lightly built and perhaps more 'generalist' early *Homo*. Compared with apes, probably all were more dependent on roots, tubers, nuts, seeds and animal protein, as confirmed by modern isotope and microscope studies, which are still exploring the details of their diets. *Homo habilis* – once thought by Leakey and Tobias to have been the first members of the genus *Homo* (but now preceded by some other candidates) – occupies an important place in the history of palaeoanthropology. It was found at Olduvai Gorge in the same levels as a robust australopithecine (*Zinjanthropus*, or properly *Australopithecus boisei*) and among the simple stone flakes and cores known as the Oldowan tradition. The maker of these tools was thought to be the more gracile and human-looking fossil, hence their Latin species name, *Homo habilis* or 'handy man'. Now we know that stone toolmaking

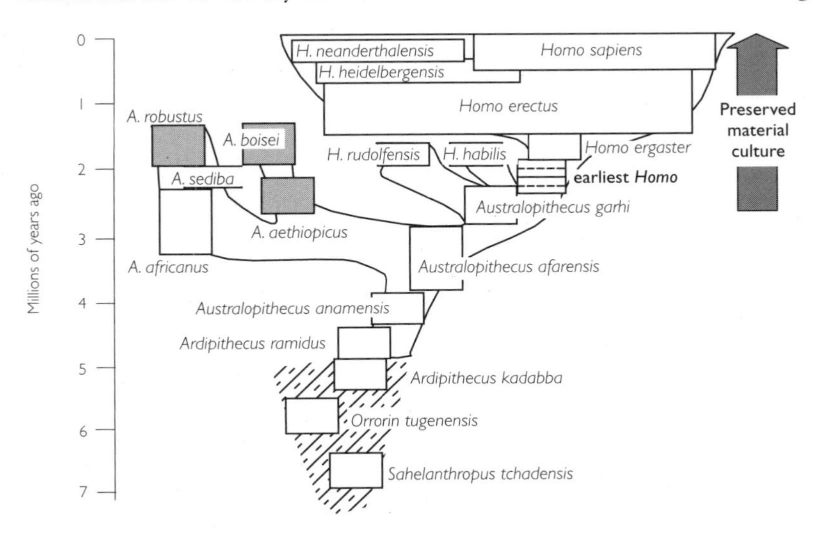

Figure 1.9: *A chart showing the main hominin species known through the last 7 million years. The relationships are much debated, but our interpretation indicates an ancestral stem and a radiation of species that began around 3 million years ago.*

began more than half a million years earlier, and which species made them is open to question. It most certainly does not have to be a member of the *Homo* lineage – after all chimpanzees are adept toolmakers and users, and we cannot be sure that any of the later australopithecines did not make stone tools. Although the roots of *Homo* are hard to discern because of a paucity of fossils between and 2.0 and 2.5 million years ago, the diversity of finds from just after this period suggests a complex early history. This was a period of rapid climatic change and was associated with many extinction and speciation events, so tracing the links is not easy, but rare finds from Ethiopia and Kenya dating to 2.3 to 2.4 million years show that some form of early *Homo* was certainly present.

By 1.9–1.8 million years ago we find a diverse group of early *Homo*, among which one species – *Homo erectus* – was highly successful as measured by its evolutionary longevity. Including the local varieties often known as *Homo ergaster* in East Africa and *Homo georgicus* in Georgia, the *H. erectus* group dominates the record of the next million years. They were probably the first hominins to escape from the confines of Africa to colonize large parts of Eurasia.

Although it is customary to distinguish between early African (*Homo ergaster*) and later Asian (*Homo erectus*) species, these are in reality part of a single group of highly successful Old World species that shows a good deal of temporal and geographical variation. *Homo erectus* developed a distinctive toolkit, focused around the handaxe, whose design and functions remained largely unchanged for a million and a half years. These first members of our genus differed from their australopithecine predecessors in their taller, more elegant body shape, and their significantly larger brains. They were clearly more nomadic, built to travel distances and even, some argue, for endurance running, which would have given them an edge in hunting. Throughout their long history we might expect to find *Homo erectus* as a restless species with small populations budding off on a regular basis and moving in either direction between Africa and Europe.

While *Homo erectus* survived in Asia until as recently as 50,000 years ago, there were changes afoot as early as 600,000 years ago in Africa. One of the African *Home ergaster* populations began to develop larger brains and evolved, perhaps via a series of short-lived intermediate species, into *Homo heidelbergensis* (named after the town of Heidelberg in Germany where the first specimen was unearthed in 1907). The toolkit of *Homo heidelbergensis* was an advance over the handaxes of *Homo erectus*, and included some of the first composite tools, where stone was bound to wooden hafts to create spears.

Homo heidelbergensis underwent further evolutionary development, gradually giving rise to the Neanderthals (*Homo neanderthalensis*) in Europe and anatomically modern humans (*Homo sapiens*) in Africa. It was not until around 60,000 years ago that anatomically modern humans left Africa and skirted the southern coasts of Asia as far as Australia. Though they had surely crossed paths with Neanderthals in the Levant, it was only around 40,000 years ago, when they peeled back into Europe from the steppes of southern Russia, that they really came into contact with these northern hominins.

The Neanderthals had survived in Europe for several hundred thousand years by then, and had evolved effective anatomical adaptations to the harsh, sometimes freezing, conditions there. They had developed a lifestyle based around close-quarters hunting of large game, everything from deer and horses to rhino and mammoths – species that while plentiful and meat-rich are nonetheless dangerous to hunt face-to-face with thrusting spears. However, a significant change in the development of tools had to await the long cultural gestation of anatomically modern humans, who appeared around 200,000 years ago. It was not until 100,000 years later that we see in Africa the first evidence for sophisticated tools and artwork such as necklaces. We have to wait a further 60,000 years until Europe, with its profusion of figurines, bone flutes, beads and cave art – the Upper Palaeolithic Revolution – catches up in the game of symbols.

The last Neanderthals died out in southern Spain perhaps less than 40,000 years ago, as the height of the last Ice Age advanced. In the end, they proved to be less successful at coping with the rigours of these northern climates than anatomically modern humans, perhaps because they were lacking in cultural versatility. By then, modern humans had colonized Australia and they stood on the brink of crossing the Bering Strait to colonize the Americas. The great 'land-grab' by modern humans was almost over. Only the remote oceans remained for this single species to settle and thus complete its global journey – although that only happened in the last 5000 years.

Working together

What palaeoanthropologists and archaeologists have achieved over the last 50 years has been to make their Pandora's box of discoveries much bigger. If we set out to be archaeological purists, we could tell an entire story from this archaeology – we could insist that only archaeology can describe and interpret the human past from its material remains. Then we could get by, perhaps, just doing a double act with the human fossils that adorn the record. But in the end this will not do. Although the archaeology is the core of the cultural record, it has always depended on a host of scholars in other fields for illumination. Geologists, environmental scientists and dating specialists all play crucial parts in building the record. Then latterly primatologists, geneticists, neuroscientists – and of course evolutionary psychologists – have all now made their contributions.

Does this interaction mean that an integrated evolutionary story has emerged? Not yet, would be our answer. First, as our colleague the palaeoanthropologist Rob Foley has often noted, there has been too little effort to bring evolutionary theory into explanations of human evolution. Archaeologists have always treated human evolution as something unique, rather than as a 'normal' product of conventional evolutionary forces – just one of many hundreds of thousands of unique species. If we try to define humans as separate, we are forgetting the evolutionary gradient of the past – and the fact that nearly always our close relatives show some of the same features. Our aim should be to show how and why we diverged from the other apes and came to be as we are, not to make ourselves totally and artificially separate.

Two generations ago thinkers such as Julian Huxley came to appreciate the great importance of what he called psychosocial evolution. Mind was the preoccupation of several of the great evolutionary biologists of his day, including Bernhard Rensch and Theodosius Dobzhansky. What was missing, when we look back, was that these scientists focused nearly always on the anatomical and behavioural adaptations that characterized species – they scarcely considered the internal dynamics, how interactions between individuals shape a society, and influence the course of its evolution. A new cycle of interactions with psychology has been necessary to highlight these forces.

For all the developments that we have explored, there has remained a gap, first clearly outlined by the French thinker Pierre Teilhard de Chardin. The psychosocial domain of Huxley was for him the noosphere, or sphere of human thought; he talked of the 'irresistible tide that for the last hundred years has been bringing natural history and human history closer together'. Even so, historians have treated social evolution as outside and separate from biology. As Teilhard de Chardin put it, 'The domain of zoology and the domain of culture: they are still two compartments, mysteriously alike, maybe, in their laws and arrangement, but nevertheless two different worlds.' Yet the one necessarily shades imperceptibly into the other.

In recent years, primatologists such as Andy Whiten and Bill McGrew, and psychologists like Michael Tomasello, have attempted to come to grips with these issues of culture. Compared with the activities of chimpanzees, of course, the permutations of human culture remain undeniably complex. As a result, socio-cultural anthropologists have often regarded the biologists' efforts as absurdly 'reductionist'. However, in doing so, they have misconstrued what the biologists are trying to do. Seen in proper perspective, biologists are exploring foundations, the underpinnings of social behaviour, not explaining why we go to the theatre, hold weddings or go to art exhibitions.

Summary

In the following pages, we try to grapple with all these conundrums. The Lucy project brought to the table two rather different ways of tackling human evolution: the psychologists introduced the perspective of an experimental science to which the archaeologists added the methods of a historical science. Bringing together the quite different spheres of archaeology and evolutionary psychology, we aim to stake out some of the main positions. We attempt through the social brain to find a better approach than the purely archaeological–material, or one based on just projecting ideas from the present. In particular, how could archaeologists benefit from perspectives provided by the social brain? Perhaps they need an impetus to escape from their intellectual paradigm – it seemed so obvious that creatures who evolved to be clever would end up on top, that the drive to be clever does not need explaining. When, in 1921, the archaeologist Osbert Crawford said, 'It may seem a far cry from the first generalized stone implement to the latest highly specialized aeroplane; but once the first step is taken, the rest is comparatively easy', we have to ask 'Yes, but why?' Why go through such an astounding series of changes? Those of us alive today are, of course, the end point for the great sweep of human evolution. From the traits we have evolved flow the capacity for everything that makes us who we are – the capacity to live in large political organizations, to engage in warfare, culture, storytelling, religion and science. At the other end of this trajectory lie the apes, about whom we also know a great deal.

Nonetheless, our task will not be to try to explain this diversity or why some species failed and others lived. This is not a story of a steady progression up the grand staircase of evolution. Our family history has been one of myriad branches that have explored any number of alternative pathways to survival. Some, like the Neanderthals and many of the australopithecines, were very successful in their time, but eventually succumbed to the vagaries of climate change and ecological competition. Rather, our task will be to explain the convoluted story of our particular species, the twists and turns that led from a perfectly ordinary ape in the forests of Africa to the species that has eventually, for better or for worse, come to dominate the planet on which we live.

The conventional way of telling this story is in terms of the succession of fossils and tools, of the anatomical and sometimes ecological differences between ancestor and descendent species. We will attempt a different approach: what does it mean to be human, and how did we come to be that way? The focus is on psychology as much as anything, and on the interplay between the cognitive and social aspects of our behaviour and the tools and artifacts that our ancestors used, and then left behind.

2

What it means to be social

The price of social life

The great evolutionary invention of the primate family is sociality. Living in groups does not come cheap, however. The more animals in the group, the further you have to travel each day because each animal has to forage in a roughly constant area to find the food it needs. That adds a burden because the animals could have been resting quietly in the shade of a bush or socializing with their 'friends'. Living in groups also incurs physiological costs: the stresses created by the times when others bump into you or try to displace you from a particularly juicy food patch or safe sleeping site. These occurrences inevitably increase with group size. Even though individual events may be quite minor, the accumulation of many of these, day after day, creates stress, and stress hormones like cortisol not only cause wear and tear on our bodies and minds, but can also be especially disruptive for females. Stress, whether physical or psychological, has the effect of destabilizing the hormones that drive the menstrual cycle: the result is amenorrhea – menstrual cycles in which ovulation does not occur, making the female temporarily infertile.

These are heavy costs to bear, and make living in groups less than worthwhile *unless* there is some other benefit. For monkeys and apes, that benefit is protection from predators. By grouping together, they make it more difficult for predators to successfully pick off individual prey. They

Figure 2.1: *Leopards are the main predators of baboons and many other Old World monkeys and apes – though baboons will defend themselves on occasions.*

may even be able to mob an attacker and drive it away. Indeed, baboons have been recorded doing this in Africa. Predators such as hyenas and leopards pose one of the main problems that primates face in their daily lives: reducing the risk of being taken unawares thus becomes a matter of – quite literally – life and death. This is a particularly salient issue when the relative safety of the forest is left for the open plains, where places to hide are few and far between. However, the prospect of prey for a would-be hunter is much higher (see Chapter 3).

The 'rule of 3' in human communities

Human groups are not all that different from those of monkeys and apes in that they consist of multiple layers of relationships that build up to create ever-larger communities. These layers go by a variety of names – band, local group, camp group – but the one we focus on here is the community, which we link to Dunbar's number of 150.

In small-scale traditional societies, these multiple layers are built up by families that combine to form bands (or overnight camp groups), and bands that combine to form communities. Membership of overnight camp groups changes over time, as families or individuals decide to come and go. But when they do switch between camp groups, they invariably do so with camp groups that belong to the same community of around 150 individuals. And as we saw in Chapter 1, this is a community whose members they already know. In foraging societies, these communities (sometimes called clans or regional groupings) are typically a group of individuals who have rights of access to particular resources, such as permanent waterholes. In settled, horticultural societies, they are usually manifested as villages that own land.

Moves between communities, in contrast, are much less common. Nonetheless, communities themselves may be banded together into larger groupings that have more congenial relationships than do strangers. Such super-communities have been referred to as 'mega-bands' in the archaeological literature or 'endogamous bands' by anthropologists, though they are not really bands in the sense of overnight camps. They are more like trading networks, in which neighbouring communities know each other well enough to be willing to trade items such as raw materials for making tools or even prepared artifacts or other items that may be difficult to make for oneself. They also serve as a network to find marriage partners.

The hierarchy continues beyond the mega-band level, with mega-bands in turn combining to form even larger groupings of individuals who speak the same language. These are sometimes referred to as tribes or ethno-linguistic communities (anthropologists often avoid the word tribe, but in this particular sense it fits well, and has frequently been used in Australia where the structures were particularly evident). These layer-like groupings turn out to have very specific sizes, and these sizes scale with respect to each other with a ratio close to three. In other words, each layer is three times larger than the one below it. Tribes are three times larger than mega-bands, mega-bands are three times larger than communities, and communities are three times larger than bands (see Table 2.1). The numbers concerned are typically about 1500, 500, 150 and 50, respectively.

We see much the same pattern of layers if, rather than looking down on the landscape from above, we look upwards from below, from the level of the individual (see Figure 1.2). It turns out that, if we ask people to list all their friends and relations and also say how often they see each of them, the pattern of relationships that emerges has exactly the same form: we are surrounded by layers of relationships that differ in both the intensity of the relationship and the frequency with which we contact the people concerned. These layers at 5, 15, 50, 150 and 500 are

Social groupings among hunter-gatherers	Numbers	Personal network
Tribes (language)	1500	Far acquaintances
Mega-bands (marriage and trade)	500	Near acquaintances
Communities (Dunbar's number)	150	Friends
Bands (overnight camp groups)	50	Good friends
Foraging group (support group)	15	Best friends
Intimate group (soul mates)	5	Close intimates

Table 2.1: *The rule of 3 in hunter-gatherer societies (top down) and personal networks that apply to all societies (bottom up).*

more or less synonymous with intimates, best friends, good friends, friends and acquaintances.

In one important respect there appears to be a big difference between the community of 150 and the layers beyond it: those within the 150 constitute a set of individuals with whom we have reciprocal relationships of trust and obligation. They are relationships that have history – we have known these people for some time, and they know us. Those who fall into the layer beyond the 150 are what we might call acquaintances – people with whom we have rather casual relationships, which don't involve obligations of reciprocity, of helping out as a matter of obligation. This radical difference dramatically affects our willingness to behave altruistically towards other individuals.

The outermost layer, the layer that extends out to include about 1500 people, seems to correspond to the number of faces we can put names to – a pure memory problem, where the limit is set only by the capacity of the brain's memory banks. This is an example of the limits imposed by the cognitive load of remembering and acting on so much social information (see Chapter 1). Among the 1500 we would list, in addition to our family, friends and acquaintances, all the people we recognize but don't have any kind of relationship with: people that we know, but who don't know us. For most people, these would no doubt include President Obama and perhaps the Queen of England, an assortment of rock stars, the regular anchor from our regular television news programme, a celebrity we follow on Twitter, and so on – people we would recognize in the street, though they wouldn't have the faintest idea who we are.

The fact that these two views – the world as a whole seen from above and the social world of the individual seen from below – coincide so closely is puzzling, and we don't have any real explanation for it, except perhaps to suggest that the world of organizations may take the form it does because organizations are based on the individual members' personal relationships. In other words, organizations have the layers and the sizes they do because they emerge out of the limits imposed by individuals' abilities to handle relationships of various intensities.

The shape of armies

The 'rule of 3' that seems to define the successive sizes of the layers is repeated in one other context in real life that is worth mentioning, and this is the military. Modern armies all have essentially the same structure, having evolved from the casual units that feudal barons brought to the defence of their overlords. During the Thirty Years War that devastated northern Europe between 1618 and 1648, Protestant and Catholic armies rampaged around the countryside creating havoc for the peasants and a great deal of death and destruction among themselves. The leading Protestant combatant was King Gustavus Adolphus V of Sweden, and his contribution to military history was to set in place the beginnings of modern military organization. The problem he faced was essentially one of management. Winning on 17th-century battlefields was a matter of solving two incompatible problems: maximising the weight of manpower on the battlefield (the bigger your army, the more likely you were to win – most of the time) while maintaining coordination between groups of soldiers (the ability to coordinate declines precipitously with the size of your army). The reforms he instituted eventually gave rise to modern military organization, which combines structural integration with robust discipline (you must obey an officer who is senior to you). It's the structural organization part that is most interesting for us.

Modern military structure follows exactly the same rule of 3 as human communities, and does so with unit sizes that are almost exactly the numbers we find in the everyday social world. Typically, three sections of about 12 men each combine to make a platoon of about 40–50 men, and three platoons combine to make a company of about 150; three companies make up a battalion (500), three battalions a regiment (1500), three regiments a brigade (5000), and three brigades a division (15,000). Of course the actual divisions are often slightly different. Different armies use different names, but the numerical values are generally similar

– companies, for example, vary only between about 120 and 180 in size across all modern armies. The company is considered the foundational unit in this structure: it is the smallest unit that can operate alone as an independent entity, and it is very much regarded as family. Note that the military give us at least two more layers above the 1500 tribe layer that we found in traditional small-scale societies. Note also that – as we draw on archaeology – it is interesting to see how the Romans experimented and always came back to the same system. The maniple, a tactical fighting unit introduced in 315 BC, consisted of three lines of 40 men; the century of the later imperial legion was a smaller unit, but in the prestigious first cohort it was raised to 160 men.

Time, friendship and kinship

There are striking differences in the frequency with which we see the individuals in each of the layers of our personal social networks. On average, we probably spend about two hours a day engaged in social interactions – that's excluding workplace interactions that we should properly consider work rather than social, and of course interactions with our doctor, lawyer, baker, etc. You might think of this as our social capital, a fixed quantity of social effort that we can invest in each of our friends and acquaintances. Of this, something like 40 per cent is devoted to the five people in our innermost circle – so each of them gets about 8 per cent of our available time. A further 20 per cent is devoted to the ten additional people in the 15 layer, with each of them getting about 2 per cent of our social capital. The 35 extra people in the 50 layer get an average of about 0.4 per cent of our time each, and the 100 additional people in the outermost layer only a quarter as much – equivalent to seeing them about once a year. The importance of time for the social life of primates is discussed in the box opposite.

In Chapter 1, we saw that there is a cognitive constraint on the size of social groups (or, at least, a cognitive constraint mediated by brain size) – what we refer to as cognitive load. However, time also plays an important role here. In our studies of personal social networks, we asked people to tell us how frequently they saw each of their friends as well as how emotionally close they felt to them. We measured emotional closeness on a very simple 1–10 scale, which, despite its simplicity, in fact correlates well with many of the other measures of emotional closeness that psychologists

The importance of time

Time is important for monkeys and apes for two reasons. One is that they have to get all their foraging and travel between food sites done within a 12-hour waking day – not least because, at a very early stage in their evolutionary history, the monkeys and apes opted for a strictly diurnal lifestyle and so have poor night vision. Increases in body size and brain size have to be compensated for by increasing the time devoted to feeding in order to ensure that sufficient extra energy and other key nutrients are ingested.

The other is the time required for social grooming. Because primates create their social relationships through grooming, and the strength of these relationships is a direct consequence of how much grooming time the animals exchange, species that live in large groups have to devote proportionately more time to grooming. The bigger the group, the more time the animals have to devote to grooming each other.

How animals allocate time to these core activities is crucial to their ability to successfully colonize a particular habitat as well as to the size of group that they can live in. Understanding exactly how different species of monkeys and apes manage their time budgets, and the climatic and environmental factors that influence these, was an important part of the Lucy project. Between them, project members Julia Lehmann and Mandy Korstjens developed a series of models of the time budgets of African monkeys and apes. It turns out that the most important factors are temperature and seasonality. Temperature is important both because it influences the quality of the food available to animals (succulent fruits tend to be found in shady forests where ground-level temperatures are cooler) and because high temperatures in the middle of the day force animals to seek shade and rest (thereby shortening their active day even further).

These models were later used by Caroline Bettridge, one of the postgraduate students on the project, to explore the time constraints faced by the australopithecines and how they coped with these. She found that, had the australopithecines been conventional apes, they would have been unable to survive in the kinds of habitats where they actually lived, mainly because their travel time requirements would have sky-rocketed. Bipedalism seems to have been a partial solution to that, in part because it is energetically more efficient and because longer legs allowed some savings on time. However, that alone would not have allowed them to occupy the habitats they lived in. A change of diet was also necessary so as to reduce the time costs of feeding. That change of diet seems to have involved increased dependence on roots and tubers that provided more concentrated feeding sources.

have used. Emotional closeness indexed in this way turns out to correlate with frequency of contact: the more often you contact someone, the more emotionally close you feel to them (see Figure 2.2). One obvious implication of this finding is that if you interact less with someone for some reason – perhaps because you move away to another town and can't see them so easily any more – then your relationship with them, as measured by your emotional closeness, is likely to deteriorate quite rapidly.

We tested this hypothesis by looking to see what actually happened to relationships over time. To do this, we needed to study a group of people who would be moving away from home, so that they would find it difficult to maintain the same frequency of interaction with the original members of their social network. For this project, run by Sam Roberts, we signed up 30 students aged 18 in their last half-year at school in exchange for free mobile phone subscriptions for a year and a half. This allowed us to build up a picture of their home social networks in the six months before they went away to university, and then to follow them through their first year at university, where there are many opportunities to make new friends and where the physical distance from home meant that they would be unable to see much of their old network.

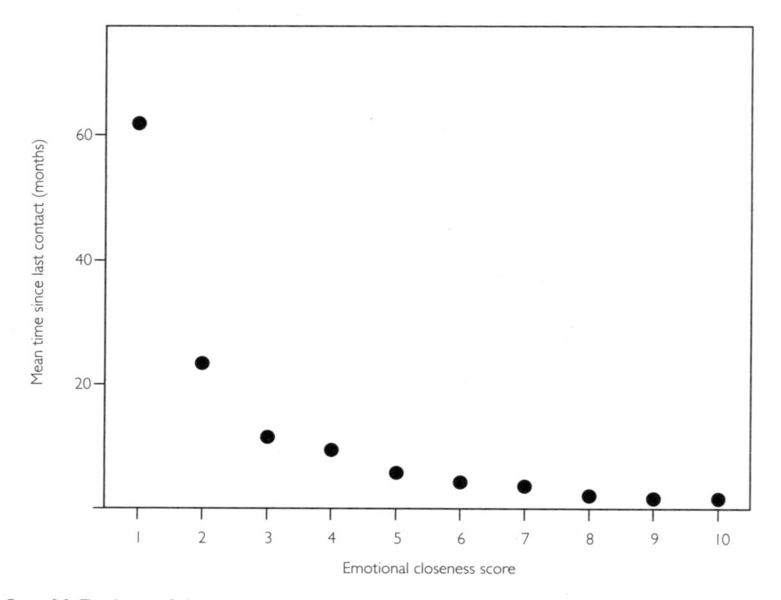

Figure 2.2: *The closer we feel to someone, the more often we see them. Here, the mean number of months since last contact is plotted for relationships of different emotional quality (indexed by the emotional closeness, where 10 is very close).*

The results were startling. When the frequency of interaction declined, the sense of emotional closeness in their relationships plummeted within a matter of six months at most. It really happened astonishingly quickly. Of course, this might just be a peculiarity of teenagers and the fickleness of teenage relationships. However, we do not think so for several reasons. First, these were not children, but people on the verge of adulthood; by the time the study ended, they were well into their twentieth year. Second, ours is not the only study to have demonstrated such effects: previous studies have shown that even adults' friends change when they move towns. Our project differed from these earlier studies in that we could specify not only which relationships died and which didn't, but exactly how this related to perceptions of emotional closeness and to the frequency of interaction. In short, time is everything, and when time is not invested in a relationship, relationships wither away.

Relationship quality is important because it affects how altruistic we are to each other. In our studies, there were quite marked differences in how altruistic we are likely to be towards our friends depending on how close they are to us. The closer they are, the more willing we are to help them or to do them a favour. In another of our studies, Oliver Curry asked subjects to nominate someone in each of the layers of their personal social networks out to the 150 layer, and then to say whether or not they would donate a kidney to them if asked. They were about 15 per cent more likely to do so for someone in the inner 15 layer than for someone in the outer 150 layer.

Of course, it is one thing to ask whether you would be willing to do something like donate a kidney, and quite another for you actually to do it on the day. But in another study, we asked subjects to do a painful skiing exercise to earn money for a named relative. The exercise was a standard skiing drill that involves sitting with your back against a wall as though on a chair, but with no chair under you. It is designed to strengthen the quad muscles so that you can improve those graceful slaloms down the piste. At first, it is quite comfortable, but it becomes excruciatingly painful after about three minutes, and most people collapse on the floor unable to stand the pain any longer after just four or five minutes. We paid subjects £1 for every minute they could hold the position, and the money they earned on each occasion was sent directly to a nominated individual – themselves, a parent or a sibling, an aunt/uncle/niece/nephew, a cousin, or a friend of the same sex. No matter which order they did these in, the amount of money earned went down in that sequence (although women – but not men – usually worked a bit harder for a friend than a cousin). (Incidentally, we also included a well-known children's charity as one of the options, and this always did much worse than any of the other beneficiaries.) This was real, genuine altruism, because subjects had to incur pain to earn money. They were willing to incur more pain for those who were very close to them than for those who were less close. Emotional closeness and altruism go hand in hand.

There is one glaring exception – friendship relationships are fragile and decay rapidly if they are not reinforced. But relationships with family members are very different. In our 18-month study of students going away to university, we discovered that family relationships were incredibly robust in the face of failure to interact. And this wasn't just relationships with parents and siblings, it was relationships with the whole extended family out to second cousins. The fact that you might not have contacted your aunt or uncle, or a cousin, for more than a year seems to make not an iota of difference to how emotionally close you feel to them, or how they feel about you. Indeed, if anything, expressed levels of emotional closeness to relatives increased with time away from home.

This kinship premium inevitably extends to behaviour. No matter where they lie in the layers of our social network, we are more generous towards relatives than towards friends. On average, kinship is worth an extra 40 per cent over friendship in terms of willingness to donate a kidney. This increased generosity towards family is nothing new, of course: it is the evolutionary process known as kin selection, whereby we favour kin over non-kin because the former share some of their genes with us. This, as we shall see, is an important feature of small-scale communities and traditional societies.

The emotional side

In the previous section, we talked about the importance of time for servicing our relationships. But time is not the only factor that is important in limiting the number of friends we can have. Our relationships are emotional things, and so some aspects of our psychology are likely to be crucial in how we manage them. There are two sides to this: one is empathy and the other is how well we understand our friends.

Emotions are very tricky things to study at the best of times, and psychologists have largely avoided them for almost a century. In part, this is probably because emotional responses seem to be (in a very simple-minded sense) right-brain phenomena. Perhaps because the language centres are based on the left side of the brain, and these don't connect up terribly well with the emotion centres, we find it surprisingly difficult to reflect on our emotional states, and cannot describe them very well in words. Since we don't have a language in which to describe these internal states, we lack any kind of scientific metric by which we can measure the intensity or quality of an emotion. And that means it is almost impossible to compare the emotional states of two different individuals. Is my sadness or happiness greater than yours, or less? We just don't know and we can't really tell – which is no doubt why teenagers think that their own distress is much worse than anyone else's could ever possibly be. It was this impasse that persuaded the behaviourists in the 1920s to argue that we should avoid all discussion of mental states in animals (and humans) and instead concentrate on what we can actually see and measure, namely *behaviour*.

We might have been tempted to duck the problem of emotions, but we soon realized we cannot. Emotions play an important role in our relationships, which are clearly associated with feelings of warmth and happiness, and many of them are terminated amid especially intense feelings of anger and upset. In one of our studies, run by Max Burton, 540 subjects told us about relationship break-ups they had experienced in the past year. A surprisingly high proportion (65 per cent) of these were with close family members (out to and including cousins), although inevitably the single largest category was romantic partners (34 per cent). The most common reason for relationship breakdown was perceived lack of caring, with jealousy and envy not far behind. All these were attended by a dramatic decline in feelings of emotional closeness, not to mention associated feelings of anger and distress.

Of course, when relationships are working, we feel the opposite kinds of emotions, though it has been hard to pinpoint just what these actually are. Our verbal descriptions of our states of mind when relationships are in full swing are often incoherent and sketchy. But there is one aspect that we can say something about, and this is the feeling of warmth we get from successful relationships.

Emotions		
Social	Guilt, shame, compassion, pride	Needs theory of mind to have any effect on behaviour
Primary	Fear, anger, happiness, sadness	Provides the response to threat and need
Mood	Haunting, enchantment	Affects a place and a social gathering

Table 2.2: *A ladder of emotions and some selected examples.*

There is another dimension to emotions that directly relates to our evolution and the social brain. As shown in Table 2.2, emotions can be grouped into three levels. At the base are the mood emotions, the vibes we pick up from places and people. We respond to feelings such as security and apprehension without necessarily being able to put our finger on exactly how they are generated. A safe haven and a spooky place are very basic to our understanding of place and people. At the next level we find the primary emotions. These make sense from a survival point of view. Fear, anger and happiness are emotions we share with all other mammals. They provide us with the emotional response to threat and danger. Finally there are the social emotions. These are more complex and include the human feelings of guilt, compassion and pride. What makes them distinctively human is the capacity for mentalizing that is needed to make them work. Shame depends on the recognition that someone else has a belief. (We are aware that dogs are also very good at guilt and shame – but Daniel Dennett, mentioned below, has seen them as the great animal exception, so moulded by us that they take on many of our states of mind.)

Intentions, mentalizing and theory of mind

We need here to say something about an aspect of the psychology of relationships that seemed at first to be wholly opaque, but which we have learned a great deal about in the last decade or so. This is the phenomenon we have referred to in Chapter 1 and above as mentalizing, the ability to understand another individual's mind state, and particularly their intentions – from whence comes another name for it coined by the philosopher Daniel Dennett, the *intentional stance*.

The intentional stance refers to our extraordinary ability to understand just what it is that someone else is trying to convey when they speak. Words are notoriously slippery things, and of themselves are often quite ambiguous in their meaning. In fact, sometimes they can even mean quite the reverse of what they seem to say, as when we use metaphors or sarcasm. We are able to work out the real meaning of a statement because the speaker usually gives us clues in their tone of voice or the gestures they use. The cognitive skill of mentalizing is the reason we can do this complex, but to us ordinary, social task. Mentalizing is related to empathy, but it differs in that empathy might be regarded as a 'hot' form of cognition (we *feel* what the other person feels) whereas mentalizing is more like 'cold' cognition (we *understand* it). It's a skill that we use on a daily basis to figure out just what it is that everybody else wants, how they might react to something we do, and how we might best act so as to get them to do what we want them to do.

Children first acquire this ability around the age of five years, when they realize for the first time that other people have minds of their own – minds that might even result in them believing something quite different about the world. Psychologists have referred to this ability as 'Theory of Mind' (meaning that the child now has a theory about the mind). Once they have mastered this skill, children can do two important things that they couldn't previously do at all well. One is to lie with conviction, because they now know how you will take what they say, thus allowing them to feed you false information. The other is to engage in true fictional play. The dollies' tea party now has reality: the dolls might spill the contents of their (empty) cups and mess their dresses. Or a piece of wood on a length of string is a real car making its way through the garden race track. This is an important capacity that sets in train the possibility of something far more important than children's games: it is what eventually makes culture possible. We return to that part of the story in a later section of this chapter. But first, how does this capacity relate to social relationships?

Theory of mind, as mastered by five-year-olds, is – in the grand scheme of things – not a massively impressive ability. Part of it involves being able to adopt another individual's perspective, and we know that perspective-taking is something that even other monkeys and apes can do. Although theory of mind is a bit more than just perspective-taking (it involves not just understanding others' perspectives, but using that to understand their intentions), theory of mind is probably something that we share with other great apes. The intentionality stance provides us with a natural metric here, since intentionality forms a hierarchy that we might think of as a series of reflexive mind states. The fact that I have a belief about something (I know my own mind) is equated with first order intentionality, and my having a belief about your belief (your state of mind) is then second order intentionality. This is the state that five-year-olds arrive at when they have mastered theory of mind. But adults can do considerably better than this. In fact, our studies suggest that fifth order intentionality is the natural upper limit for the majority of people. This is something equivalent to being able to say: I *wonder* whether you *suppose* that I *intend* that you *think* that I *believe* X to be true. The five words in italics are the mental state terms that are collectively referred to by philosophers as *intentionality*.

That's pretty impressive, but some people can do even better than that. Our research suggests that about 20 per cent of the population can cope reliably with sixth order intentional statements, and a tiny handful can even manage seventh order ones. Of course, there is variation in the other direction to. A modest proportion of people can only manage fourth order, and a few probably get stuck after third order. The tails of this distribution are very long, because a small number of adults fail even to master theory of mind (second order intentionality); these people are usually defined clinically as autistic, and essentially only have first order intentionality. They provide a particularly poignant yet instructive example, because this deficit is associated with a complete failure to cope well with the adult social world. This is so even in those cases where the individuals concerned have normal (or even above normal) IQ.

Since these mentalizing skills are so essential to our ability to navigate our way through our incredibly complex social world, we wondered whether people's competences on these intentionality tasks might be related to the size of their social circles. In a way, there is an analogy with juggling: just as skilful jugglers can keep more balls in the air at the same time than less skilful ones, so someone who can manage sixth order intentionality might be able to keep more people in their social circle than someone who can only cope with fourth order. This was first tested in the project by Jamie Stiller. He used a series of very short (approximately 200-word) stories or vignettes to test people's abilities to understand the various mental state events in the story. There was a highly significant correlation between peoples' competences on these tasks and the number of friends they listed for the inner two layers of their network (the 15 layer; see Table 2.1).

How our brains embody social behaviour

The social brain hypothesis tells us that, across primates as a whole, social group size is limited by neocortex size, or at least by some aspect of neocortex size. An obvious extension of this claim is that, if it holds between species, then it ought to hold within species too. In other words, we should be able to show that individual differences in brain size correlate with differences in social network size. This was such an obvious prediction that, along with our colleagues Penny Lewis, Joanne Powell and Neil Roberts, we set about testing it directly using powerful new brain-scanning technology. Brain imaging machines are capable of creating pictures of the brains of living people through the skull by picking up the different densities of matter or by the tiny signals of electrical activity or blood flow in the brain. Different techniques allow you to do different things – one allows you to measure the volumes of different parts of the brain, while another allows you to see how hard the brain is working when it is doing a particular task.

The tests conducted were fairly straightforward in that all we did was ask our subjects to list all the people they had been in regular contact with during the previous month (roughly equivalent to the 15 layer) and then run them through the brain scanner while they were doing the same intentionality tests that we talked about in the previous section. We found two important things. First, that the same areas of the brain light up when subjects are doing these multi-level intentionality tasks as light up when people are doing simple theory of mind (i.e. second order intentionality) tasks. These include locations in the temporal lobe just above the ears and some parts of the prefrontal cortex just above the eyes (see Figure 2.3). However, the novel finding from our study was that these areas are bigger in people who can cope with higher orders of intentionality. Second, the orbitofrontal cortex, in particular, is bigger in people who have more friends and can work at higher orders of intentionality. More importantly, it turns out that there is an explicit causal relationship here: people who have bigger orbitofrontal cortices are able to cope with

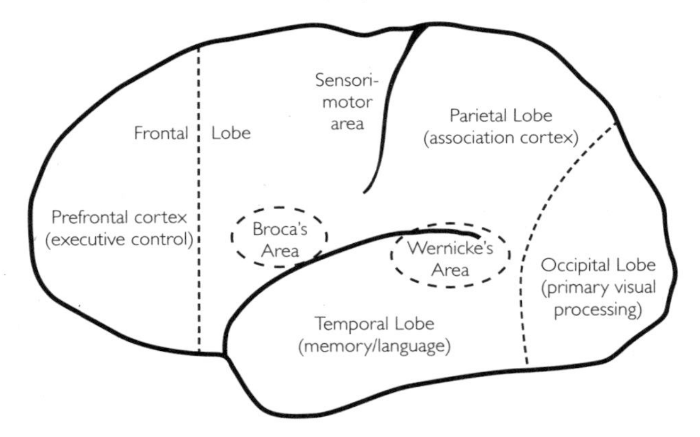

Figure 2.3: *The human brain, with its major divisions and their functions.*

higher orders of intentionality, and because they can cope with higher intentionality levels they have more friends.

This rather amazing finding tells us two things. First, the social brain hypothesis works within species just as much as it does between species. That is comforting, because it provides the evolutionary ratchet that allows selection to increase brain size as species evolve. Evolution occurs when natural selection acts on differences between individuals because those differences are correlated with fitness (essentially the number of descendants you leave). So we can hypothesize that those individuals with bigger brains (or at least bigger orbitofrontal cortices) were more successful socially, and so left more children and grandchildren, resulting in a steady increase in brain size across the generations. We don't know this for certain for humans, but we do know from studies of baboons in East Africa that females that have more friends (defined in the absence of language by their regular grooming partners), and have more surviving offspring than less social females do. Friends really do matter, and it is probably because they buffer you against the irritating stresses of living in groups. They come to your aid when you are attacked, and, perhaps more importantly, their mere presence near you keeps thugs away. Second, it reminds us why the social world can be so difficult to deal with. It takes a great deal of brainpower, even though the process is implicit (i.e. instinctive) rather an explicit (i.e. involving conscious thought).

And that should remind us of a further evolutionary point. Since brain tissue is extremely expensive (it consumes about 20 times more energy per gram than muscle tissue), evolving a bigger brain does not come for free and so must have a very strong selection pressure in its favour. The social brain hypothesis tells us exactly what that selection pressure was: the need to create large coherent social groups that involve a proportionately large number of relationships. However, it does not, of course, tell us what the selection pressure was that demanded the large social groups, although we know from studies of monkeys and apes that the need to keep predators at bay seems to be the important factor. What we can infer is that the pressure from predators was sufficiently intense that it forced those that wanted to colonize dangerous habitats to develop a social counter-strategy. We will see later that this doesn't entirely explain what happened in the human lineage, where other forces eventually seem to have come into play. But it reminds us that we cannot ignore the fact that there is a question lurking beneath any relationship between brain and cognition that begs an evolutionary answer. Since brains don't come for free and animals have to feed both more and better to pay for big brains, they need a seriously good reason to justify going down that evolutionary pathway.

Grooming and the chemistry of the brain

We earlier raised the thorny question of emotions and how difficult they have been to study. We now need to return to this topic to examine another important aspect of social bonding. In monkeys, apes and humans, social bonding seems to involve some kind of dual-process mechanism. One part of that mechanism is the cognition that makes it possible for us to do the mental calculations involved in processes like theory of mind. It also underpins the more explicit cognitive processes involved in trust and reciprocity, where we keep track of who does us favours and who reneges on their obligations. The other part of the mechanism, however, is very different and is more akin to the forms of hot cognition that we associate with emotions. It arises from the fact that a class of neuropeptides known as endorphins, and in particular a sub-class of these known as β-endorphins, seem to be intimately involved in social behaviour in monkeys and apes. These chemicals are produced in the hypothalamus (a tiny area buried deep in the old brain below the neocortex), but their receptors are widely distributed throughout the brain, although they are especially dense in areas associated with managing pain.

The important thing about endorphins is that they are released by the brain in response to pain or stress on the body. Even psychological stress will release them. Because they are chemically related to morphine, they numb the pain and give us the same uplift that morphine and other opiates do. The only difference is that we don't get physically addicted to the brain's own version of this family of chemicals in the way we do to artificial opiates. Social grooming is one of the contexts in which endorphins are released, and social grooming is, of course, the central mechanism involved in social bonding in monkeys and apes (Figure 2.4). This is because of a special set of nerves that respond only to light touch and the kind of stroking of the fur or skin that is involved in grooming, triggering the release of endorphins in the brain.

Opiates are rewarding, which is why people keep going back for more. The same is, naturally, true of endorphins. The pleasure we get from them makes us want to repeat whatever we did that triggered their release. At the same time, the relaxing effect that they have creates a psychological frame of mind that allows us to build a relationship of trust with whoever we happen to be doing it with. Monkeys and apes do not groom at random with everyone in their group. Rather, they have very specific grooming partners who also act as crucial allies, protecting them from the inevitable stresses of group-living.

Physical exercise triggers the release of endorphins as a natural consequence of the pain created by the activity, and we end an exercise session with an opiate 'high' and a feeling that all is well with the world. Indeed, many of us choose to exercise regularly, as a casual walk past any city-centre gym at 8 o'clock in the morning will amply confirm. And we certainly do get all the rewards and benefits on offer from the opiates triggered. Indeed, it seems that endorphins help 'tune' the immune system and give us real medical benefits. However, there seems to be an added punch that comes from doing these activities socially. Doing an endorphin-releasing activity with someone else ramps up the effect and somehow makes it much stronger. We don't understand why, but its consequences are plain to see in monkeys and apes: grooming each other leads to very supportive coalitionary relationships that result in grooming partners being willing to defend each other, if

Figure 2.4: *Being groomed is very relaxing for this Japanese macaque, thanks to the endorphins triggered by the stroking actions of the groomer.*

necessary against overwhelming odds. In a word, endorphins create friendships and build relationships.

There are at least two other circumstances that turn out to be major triggers for endorphin release. These are laughter and music. We were intrigued by the fact that both of these peculiarly humans activities are *so* enjoyable for us – so much so, in fact, that we are willing to spend significant sums of money to experience them. Although the psychologist Steven Pinker famously dismissed music as being mere evolutionary cheesecake of no particular consequence or value, there is a rule of thumb in evolutionary biology which says that if an organism is willing to invest heavily in something, it cannot be functionless. So the fact that we are happy to spend money for something should alert us to the likelihood that it has real functional value in evolutionary terms. We shall return to exactly what this benefit is in Chapter 5. For now, we just want to establish that music and laughter are involved in the endorphin story.

To explore this, we ran a series of experiments on both laughter and music using pain thresholds as a proxy for endorphin release (as also used in research on pain). The logic is very simple. If endorphins are part of the pain control system (and, explicitly, are released when pain is experienced), then an increase in pain threshold following a bout of laughter or music-making is evidence of endorphin release. With this in mind, our experiment measured tolerance to pain by slipping a jacket for chilling wine bottles around the subject's arm. Their pain threshold was measured by how long, or short, they could put up with it. Then the subjects were given an activity that made them laugh or involved performing music in some way, after which we measured their pain threshold again. If there was no change in pain threshold, or the threshold was *lower* after the activity than before, then we could conclude that there was no endorphin effect. But if pain threshold was *higher* after the activity, then it must be because endorphin activation has occurred. To be sure of this, alongside the experimental group we also ran a control group who performed a similar activity but one that did not involve laughter or active music-making. With an increase in pain threshold in the experimental condition, but no change in threshold in the control condition, we could be certain that the activity in question was triggering endorphin release.

We ran a total of six series of laughter experiments in all, including one at stand-up comedy events at the Edinburgh Festival Fringe. Except for this last case, we used commercially available comedy videos to make subjects laugh, and compared them with subjects who watched what we hoped were very boring videos such as tourist promotion films, religious programmes or golfing instruction videos (they can be *very* dull!). The music experiments were slightly different, in that we needed subjects to actively perform music. We ran three series of experiments in this case. One compared two religious services (a prayer meeting with an all-singing-and-dancing charismatic service); a second compared a drumming circle with the floor staff from a major music retailer (who listened to music a great deal all day, but didn't actually perform it); and a third looked at 'flowing' versus interrupted musical performance (as in rehearsals where the music is constantly being stopped to correct errors).

The results of all of these experiments were essentially the same: subjects in the experimental groups who laughed while watching comedy videos or who actively performed music all had elevated pain thresholds after their respective tasks. But subjects in the control conditions, whether they watched boring videos, prayed or just listened to music, did not. In other words, laughter and musical performance are good mechanisms for triggering an endorphin effect. And they do this because they are stressful for the body.

This is perhaps more obvious in the case of playing an instrument: it is a very physical activity and involves a great deal of effort (and sometimes psychological concentration, which is also stressful). Laughter and singing are probably stressful for the body for a slightly different reason, namely the fact that the diaphragm and chest-wall muscles have to work hard and in a very controlled way to produce the sounds we require. Singing is much harder work than speaking, for example. As well as being hard work for the muscles, laughter is exhausting because it involves a series of exhalations without drawing in breath, which empties the lungs and leaves us breathless. It is not for nothing that we speak of 'laughing till it hurts'.

Although we share laughter with the chimpanzees, and probably the other great apes, human laughter is structurally rather different. When chimpanzees laugh, it is a simple series of exhalation/inhalation pairs: each laugh is followed by a drawing in of breath. This key difference means that apes don't suffer the same exhaustion effect that we do. What humans seem to have done is take the basic ape laugh and remould it in two important ways – structurally, so that it becomes more tiring, and socially so as to create the enhanced endorphin effect that we experience with other social endorphin releasers like grooming.

Musical performance seems to act in much the same way – and is, of course, very social as well. It is important to be clear that it is active *performance* of music that works in this way: listening doesn't have anything like the same effect, although whether the emotional wrenching that music can create in us is related to the endorphin effect or that of other neurological or neuropeptide mechanisms is an open question at the moment. But, we suspect that this effect also explains why we enjoy dancing so much, and why it plays such a central role in human social behaviour.

Summary

In this chapter, we have focused on aspects of modern human behaviour – the end point of our evolutionary history. We have found a remarkable fit between the structure of small-world populations such as hunters and gatherers and the levels we can find in our own social networks. These groupings follow a rule of 3, where each level is three times larger than the one below. We have also seen how the structure of our brains imposes a limit on remembering information about people and acting on it. The burden of cognitive load works for both individuals (something you can check for yourself by looking at the address book on your cellphone and classifying who is there by the descriptions in Table 2.1) as well as for those levels in social units as varied as hunter-gatherers or armies. What we have also done is establish the importance of cold cognition; the ability to mentalize, to understand what people want and how to use such information. This led us to theory of mind and the emotional basis of much of what we do.

Already in this overview of modern social life we can begin to see a gulf opening up between the present and the past. How can we study emotions from a pile of rocks? Will we ever know what social intentions a Neanderthal might have had? In the next chapter, we step back into deep-history and ask how what we see in modern humans came to be. It is time to bring in the ancestors.

3

Ancient social lives

Digging for ideas

The hardest part for any research project is testing ideas. It is also the most fun. Archaeologists, who take over the story in this chapter, test ideas with hard evidence. We excavate to answer questions rather than to recover ancient stuff for its own sake – if we don't, then we shouldn't call ourselves archaeologists. We never stick a spade in the ground without having some idea of what we will find, or hope to find, and how this will shed light on the human story. There are clues on the surface to help us decide where to look: struck flints, animal bones and indeed earlier work that went deeper and faster than modern excavations. But even so the golden rule for an archaeologist is 'expect the unexpected'. When, as a young researcher, John Gowlett was digging at the Kenyan site of Chesowanja, he was seeking ancient stone artifacts. What he found was some of the earliest evidence for fire. And imagine the surprise in 2003 when Australian archaeologist Mike Morwood unearthed in the Liang Bua Cave on Flores, Indonesia, a set of tiny human skeletons – the so-called hobbits that have rewritten the textbooks on human diversity.

But the one thing archaeologists know they are *never* going to find are fossilized examples of friendship, kinship or fifth order intentionality as described in the last chapter. There simply isn't a stone tool that unequivocally states 'She was my friend', or a burnt animal bone that allows us to reconstruct some precise social reasoning such as 'His intention in cooking a steak was to make her believe he loved her for her mind.' Even finding a carved ivory figurine in an excavation doesn't shout out 'They now had language!' And in the same way, unearthing the oldest bird-bone flute from the German cave site of Geissenklösterle doesn't mean that we have found the moment, 40,000 years ago, when music began.

There are two common responses among archaeologists to these challenges. The first is to treat everything in the previous chapter about social behaviour as speculation, ideas and models of how the world works that cannot be tested by archaeological evidence. As a result such archaeologists retreat into their comfort zone. They describe the flints they find, refine the chronologies to organize them and stick to the fundamentals of existence – what was eaten, how it was captured and why was it chosen. The second is to accept the tough part of doing research, as we did in the Lucy project, and realize that if archaeology is to have a voice in the story of human evolution then it needs to find ways to test the ideas that flow so freely from the study of modern primates and people. Our starting point was the concept of the social brain. The challenge, explored in this chapter, is to find ways to get inside it and make it accessible to testing with archaeological data.

Big brains but little social life

Some years ago Clive Gamble wrote in *The Palaeolithic Societies of Europe* (1999) that it was long overdue to move away from 'stomach-led and brain-dead explanations [of the period] and instead direct our analytical ingenuity to the richness of the social data our evidence contains'. This was the manifesto for the archaeological component of the Lucy project. You only have to read most of the earlier accounts of Old Stone Age archaeology to realize that archaeologists hadn't come very far. There you will find accounts of Neanderthals, people who had brains similar in size to ours, and yet their social lives were seen as smaller than a troop of modern-day baboons. Compare the hugely influential book by Dorothy Cheney and Robert Seyfarth published in 1990, *How Monkeys See the World: Inside the Mind of Another Species*, with archaeologist Paul Mellars' account of *The Neanderthal Legacy* (1996). The richness of the social lives and daily interactions of a small-brained primate contrasts sharply with the almost non-social existence of a massively encephalized fossil ancestor. But Mellars at least devotes a chapter to the social life of his Neanderthals. This is a good deal better than Gordon Childe, who shaped the subsequent fifty years of such discussions in his small book *Social Evolution* (1951).

As an Australian, Childe hailed from a continent of hunters and gatherers, and as a Marxist historian he held strong views about the social forces of production and how these drew on a material base. He concentrated his immense analytical ingenuity on the origins of agriculture and the rise of civilization, and he walked the interface between human prehistory in Europe and the dawn of literate civilization in the Near East. But sadly he had a very low opinion of hunters and gatherers. According to Childe, these peoples lived at the stage of savagery, the lowest rung in the evolutionary ladder adapted from the American Lewis Henry Morgan's *Ancient Society* (1877). This was a social stage we had escaped from thanks to a Neolithic revolution that brought farming – a settled way of life for which Childe argued we should be eternally grateful.

Childe was interested in the evidence for social change over the long term. The monuments and metallurgy of prehistoric Europe, the ziggurats and royal tombs of the Near East and the towns and cities that grew from an agricultural base, these could all be spun into a powerful story of what happened in history. This was not the case when it came to the Palaeolithic evidence for those earlier hunters and gatherers. In Childe's words, 'The archaeological record is found to be regrettably but not surprisingly deficient in indications of the social organisation or lack of it in lower Palaeolithic hordes. From the scraps available no generalisations are permissible.'

The gut feeling of most archaeologists has been to follow Childe and insist that social behaviour must be seen through direct evidence. Since 1951 there has been a flowering of social approaches among all those studying communities dependent on farming. Archaeologists have produced remarkable insights into the growth of social complexity in all the continents of the world and across the far-flung archipelagos of Oceania. Social life has been analyzed in terms of evolving social structure: egalitarian, big-men, chiefly, ranked and stratified societies. The ideological and symbolic frameworks for power have been minutely picked over at demographic and social scales ranging from the household to the city state and beyond to empire. Cemeteries have been analyzed and grave goods put forward as markers of the different types of society. Evidence for warfare, armies and slaves has been sought to differentiate the power structures and labour forces available. Networks of exchange and tribute have added further detail as well as explanations of how core areas transformed their peripheral hinterlands in early examples of regionalization.

But within all this ferment of new archaeological enquiry the groups that received less attention were the Palaeolithic hunters and gatherers. The evolutionary schemes did move on from assigning them to a stage of savagery. Now they were classified as Band societies in the classic progression of Band to Tribe to Chiefdom to State of anthropologists Ellman Service (1962) and a young Marshall Sahlins (1963) (these terms are defined differently from our discussion of the rule of 3 – see Table 2.1 – where some of the terms are the same). The turning point came with the highly influential *Man the Hunter* conference in 1966. This brought together anthropologists and archaeologists and presented a global overview.

Subsequently Band societies were differentiated into simple and complex hunters and gatherers who practised different controls over the pace at which food supplies were consumed. These were described by anthropologist James Woodburn as immediate (simple) and delayed (complex) return systems. These distinctions drew heavily on modern examples and have recently been discussed in detail by one of the Lucy project's associate researchers, Alan Barnard. When it came to applying these models to the archaeological record, those with burials, art and architecture, such as the Upper Palaeolithic of Europe, became examples of complex societies while the rest, by default, were simple. These included the Neanderthals and all other fossil ancestors in the Old World.

Archaeologists remain suspicious of those who claim they can make pronouncements on the social aspects of the early human story. But the time has finally come to confront the downside of this position: that whole swathes of our history as a species have been left uninvestigated. One example is provided in the box overleaf, where monogamy is explored as an evolutionary question and where exciting new research shows, *contra* Childe, that evidence is available.

The price we pay if we give in to the pessimists is that we will never really understand our evolution because, as with all monkeys and apes, our evolutionary success has depended on our behaviour, and especially our social behaviour. The challenge for the Lucy project was to find a way around this impasse in order to be able to say something constructive and meaningful about the social and mental lives of our ancestors.

What we understand as social behaviour

What would a Palaeolithic society have looked like? To answer this question, our first step was to reject the Band model, because it doesn't tell us very much; it is a label and nothing more. But it does tell us what we shouldn't be looking for. What it enshrines is a model of society organized by looking from the top downwards (other examples were mentioned in Chapter 2). This produces a view of society as an institution, one familiar to an American citizen or a season ticket holder of Manchester United. Institutions exist before we do, rather like Prince George being born into the Royal Family and the British Monarchy. We enter into them either because of birth, like Prince George, or by choice when we pay our way to become a member. The hunter-gatherer Band was an institution designed by anthropologists to fill a void. Curiously, it is defined more by what it doesn't have – farming, towns and monumental architecture – than what it does possess. It is a small beginning indeed for the future elaboration of society. In that respect Childe was absolutely right: the 'scraps' that survive don't allow us to say very much about social behaviour. If we are looking for the hard evidence of institutions as a means to study Palaeolithic societies then we will never find them.

But there is another route to investigating social life in the past, one appropriately described, for a primate-inspired perspective, as bottom-up rather than top-down. Instead of the rules of social behaviour being determined and passed on by a social institution they are worked out by the individual. The key words here are interaction and bonding. To be social is to associate and that implies interaction between members of a community, be they baboons, Neanderthals or your Facebook friends that add up to Dunbar's number of 150.

Those interactions differ according to the resources the species bring to the business of social bonding. There are two that stand out as central to our social core, materials and the senses. The former consists of food, water, stone, wood and everything else in the environment. The latter is common to the creation of all social bonds and is due to a distant but

Sex and the monogamous brain

We focus in this chapter on the way social communities are formed. But there is one important aspect of human behaviour that is somewhat tangential to this, yet plays a key role in the business of evolution, and that is mating behaviour. A species' mating system is a fundamental component of its biology, since, ultimately, it is reproduction that drives evolutionary change. The mating system is also embedded within the species' social system. Primate mating systems are conventionally divided into a few simple types: monogamous pair-bonds, harem polygyny (in which one male lives with, and monopolizes matings with, a group of females) and multi-male polygyny (in which several males live with a group of females and compete to mate with individual females as and when they come into oestrus). The precursor to these social states is generally accepted to be a form of semi-solitary sociality in which males monopolize territories within which females range individually.

Pair-bonding, or monogamy, has been a topic of perennial interest because humans appear to form such relationships. It has always been seen as being central to human evolution on the grounds that it relates to biparental care: a man and a woman combine forces to invest jointly in feeding their big-brained offspring. Indeed, pair-bonding was seen as essential to the production of human offspring precisely because these are so expensive to rear. In fact, the same argument also underpinned the division of labour, one of the other defining social traits that are thought to characterize humans: the male hunts to provide food for his wife, while she gets on with the business of rearing their costly offspring – the ultimate cooperative economic venture.

Whether humans are, in fact, monogamous or polygamous has remained a hotly debated issue. Many argue that humans are at root polygamous, with monogamy being a socially or economically enforced condition. Others have argued that males are essentially superfluous: it is maternal grandmothers that really do all the hard work in helping women to rear their offspring (see Chapter 4), with hunting being relegated to an ancillary status as a form of mate advertisement: hunting big game is a risky business, so success in this domain is an un-cheatable marker of a male's genetic quality – the well-tuned body, the perfect mind capable of deft responses to the prey's every attempt at evasion. In fact, on most anatomical measures of mating system (such as relative testis size), humans turn out to lie exactly half way between monogamous primates (like the gibbons) and polygamous ones (like baboons and chimpanzees). In polygamous species, the males have very large testes for their body size, whereas the males of monogamous species have very small testicles. This is thought to be because males in polygamous species have to compete with each other for matings, and the more sperm they can inject on each copulation, the more likely they are to fertilize the female. Since ejaculate volume is a simple function of testis size, males who want to maximize their chances

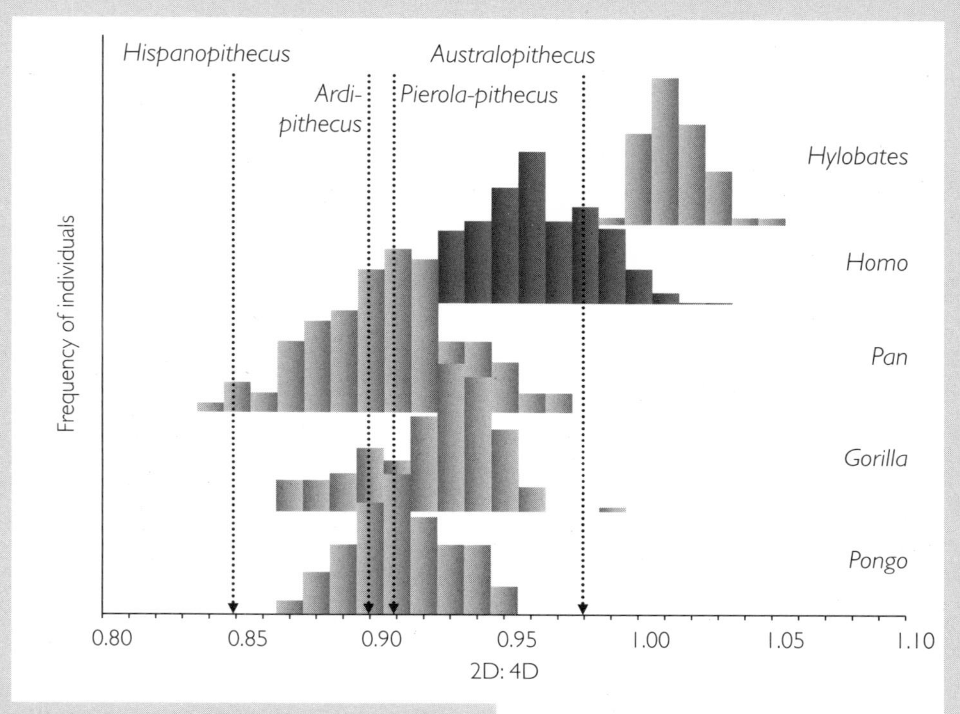

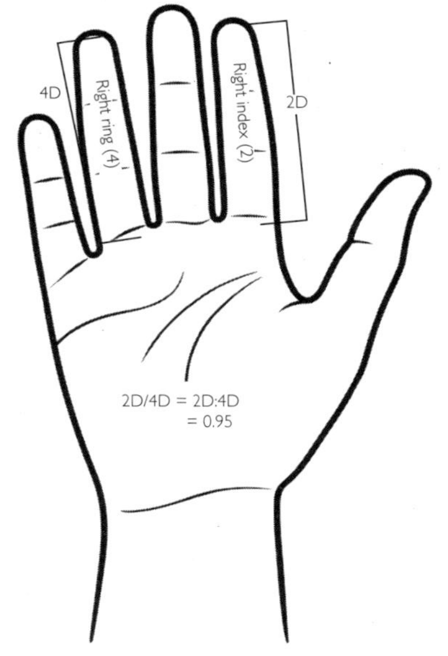

Figure 3.1: *The ratio of the lengths of the second and fourth digits (2D:4D ratio) varies between ape and human species according to their mating system. Modern humans (and perhaps* Australopithecus afarensis*) fall rather ambiguously midway between the strictly monogamous gibbons and the polygamous great apes. The Miocene apes,* Hispanopithecus *and* Pierolapithecus*, and the early hominin* Ardipithecus*, are clearly polygamous like the other great apes.*

in such a mating lottery need to have the largest testicles they can manage.

The problem, of course, is that testicles are soft tissue and don't fossilize, making it difficult to draw any firm conclusions about the mating systems of fossil species. However, Emma Nelson, a postgraduate student on the Lucy project, had the novel idea of using the ratio of the first to third fingers (usually known as the 2D:4D ratio) as a proxy. This ratio is known to be under the control of foetal testosterone, with males invariably having lower ratios than females (i.e. first fingers that are shorter than their ring fingers). She was able to show that both the males and females of monogamous species of monkeys and apes had 2D:4D ratios close to (or even slightly above) unity, whereas the males and females of polygamous species had significantly lower ratios (typically around 0.9). Inevitably, modern humans lie unhelpfully half way between the monogamous gibbons and the promiscuous great apes. In collaboration with biologist Susanne Shultz and Lisa Cashmore (another student on the project), she then examined the finger bones of fossil hominins. It turned out that most hominins (including the very early hominin *Ardipithecus* and the later Neanderthals) had 2D:4D ratios that were uncontroversially in the polygamous range. Only *Australopithecus*, paradoxically, had a ratio that was even close to that of the monogamous gibbons – and even they did not really differ significantly from modern humans with our rather ambiguous mating system.

Some archaeologists have for a long time argued that pair-bonding evolved early in hominin evolution, possibly as early as the very old *Ardipithecus* that lived between 5.5 and 4.5 million years ago – about a million years before the classic australopithecines like Lucy and her kind. But the 2D:4D data suggest that all the hominins, including both Neanderthals and even the early anatomically modern human specimen from the Middle Eastern site of Qafzeh, were as polygamous as chimpanzees and gorillas. In short, it seems that all our ancestors were polygamous. None shows any tendency to have been obligate monogamists in the way gibbons are, and few were even as monogamous as we are.

This somewhat undermines the claim that human pair-bonding evolved to enable biparental care. To the extent that biparental care occurs at all – and this is not to say that it doesn't happen – it is likely to have been a consequence of the opportunities offered by pair-bonding rather than the cause of pair-bonding, as many have previously assumed.

This perhaps makes sense, because as part of the Lucy project Suzanne Shultz, Kit Opie and Quentin Atkinson were able to show that monogamy is a kind of sink state in primate social evolution: once a species switches into it, it is very rare that it switches back out. There seems to be something final about the state of obligate monogamy. This is probably because true monogamy of this kind requires major changes in behaviour and cognition (and hence the brain) from which it is very difficult to backtrack. Obligate monogamy is an evolutionary dead end – it seems to make a species less flexible, at least as far as its social system is concerned. And loss of flexibility is not a recipe for evolutionary success. If so, it may well be that the reason there do not seem to be any genuinely monogamous hominins is that any such experiments simply resulted in rapid extinction under the conditions of continuing climate change that have dogged the past 2 million years.

common sensory heritage: touch, sight, taste, hearing and smell. And of course what all primates bring, and all hominins brought, to a social encounter are their bodies. These have the chemical pain and reward systems described in Chapter 2 as well as differently shaped digits (essential for fingertip grooming), limbs and teeth.

Then the business of social interaction and bonding can begin. The basic social unit is the dyad: a social unit of any two people, but most important for obvious evolutionary reasons a mother and child. Beyond the dyad are larger groupings of close kin, intimates, support groups and diverse networks which an individual can call on either to find food or for protection if a fight breaks out. These form levels in what Clive Gamble originally called intimate, effective and extended networks, and upon which Sam Roberts has expanded.

What distinguishes the baboon from the Neanderthal is not just brain size but also the range of materials and senses brought to the business of making the social bonds bind. With baboons this is done to a large extent by fingertip grooming. As we saw in Chapter 2, there is enough time in the day to reap the chemical rewards for such tactile bonding. Not so for large-brained hominins like Neanderthals and ourselves, with those 150-size communities. There was simply not enough time for the predicted community sizes to affirm the bonds that bind in such a close-contact way.

What took place during human evolution built on this primate pattern of working out and continually repeating who you are and what sort of society you live in through daily face-to-face interactions. It bears no resemblance to the institutional model of society with its demarcated religious, political, legal and commercial spheres. Instead it was around the core of materials and senses that humans erected the scaffolding that allowed ever more complex social bonds to be built. That scaffolding took many social forms. It included ceremonies that linked together people, both living and dead, and places both real and imagined. It involved conceiving new artifacts and architecture to express these concepts. The inspiration for these new social forms was, however, always inspired by the basic business of being social. And that meant experimenting with ways to build social bonds that bound people together at all levels of their networks and all society in more durable ways. The scaffolding has never been taken away and new social forms are always being developed – just think of our digital era. The issues of social life that we discussed in Chapter 1 still remain with us and our ingenuity has always been directed towards their resolution.

The process can be described as *amplifying* what already existed – boosting the existing signal from emotions and moods either by physical action such as dancing to release opiates or through feelings for objects, the memories they conjure up and the emotional effect, the hot cognition of empathy, they engender in the onlooker. This is easiest to see with materials. A wooden stick for teasing termites out of their mounds is elaborated into a digging stick, then a spear, an arrow, a writing tool and with each transformation new and different social relationships are built. These are the social forms of Figure 3.2. But at the same time the senses can also be amplified. Music, language, cooking and painting all release sensory stimuli that enrich social gatherings. One of the most evocative words we have found to describe this process is *effervescence*. This was used by the pioneering French sociologist Emile Durkheim to describe the buzz that social life creates, one of the rewards of coming together and performing dances, rituals and ceremonies. Effervescence also describes that lingering sensation after the gathering has dispersed, the collective spirit of association, the glue of social life.

Humans have explored many ways to heighten moods and alter the ambience of social gatherings. When we drink alcohol, listen to music and experiment with the tastes and smells of cuisines we are amplifying those core resources that bind social bonds. In the same way we engage in communal activities that are known to reward our bodies

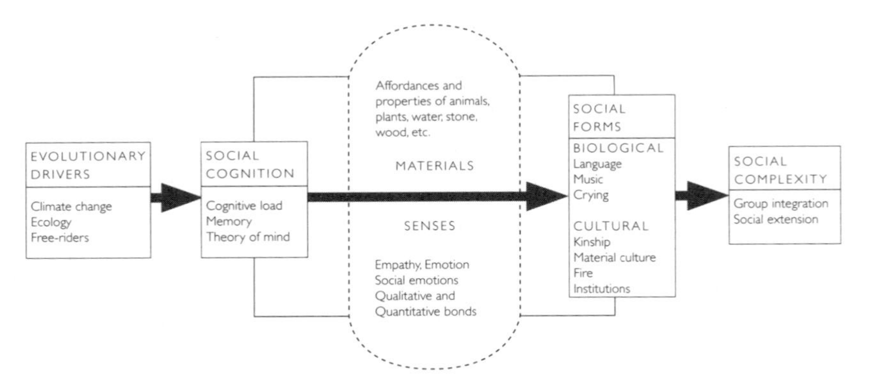

Figure 3.2: *A 'map' of the main elements involved in the social brain. The senses and materials form a core of resources that are used to make new and stronger bonds to cope with larger communities.*

with the release of opiates. In part, as we saw in Chapter 2, that explains our passion for synchronized sports, communal singing and laughing together as an audience. We also deliberately alter our immediate environment so that the place we are in either enchants or haunts while making us feel safe or scared.

Our idea of what a social archaeology looks like for the Palaeolithic therefore starts with those ubiquitous core resources as the building blocks for change. It is a bottom-up view of at least 2.6 million years of human evolution assisted by tools. Social life for every ancestor who lived in this huge timespan has been about individuals building outcomes beneficial to themselves and their offspring. They did this on a face-to-face basis and there were limits to what those outcomes could be. These were imposed by the tiny population sizes and the dampening effect this had, as argued by Stephen Shennan in his ground-breaking book *Genes, Memes and Human History* (2001), on the acceptance of novelty. In our project these issues were addressed by Garry Runciman in *The Theory of Cultural and Natural Selection* (2009).

We can be quite clear about one thing. There was never a top-down institution called society. So there is nothing for archaeologists to find in the sense proposed by Childe. Instead, the idea of society we are digging for in this book requires different ways of looking at the archaeological evidence: an examination that only makes sense with the insights of how the social brain works. Our answer to Childe's dismissal of the social organization of Palaeolithic people or other archaeologists' concerns about lack of direct evidence is simple: change the model of what you understand by social life. Forget the search for institutions that never existed. Get back to the bonds that bind and the means available to change their strength and quality. Open the social brain and look inside for its deep-history.

What we will find there has been described by John Gowlett as the eternal triangle of hominin evolution, where the three points are diet change, social collaboration and detailed environmental knowledge. Here several things in the evolutionary story come together in a crucial way. We know that early hominins needed to forage over larger ranges than their ape cousins, because of more arid and seasonally varying environments, and that the transport of stone tools proves this. But alongside this development hominins were struggling to stick together through their networks, because they needed stronger communications both for mutual support and for sharing knowledge. Put another way, the larger notional groups needed an increase in food and this also resulted in larger areas, or home ranges, for people to cover.

As part of the Lucy project, James Steele and Clive Gamble have examined ape- and carnivore-scale ranges, while Susan Anton and her colleagues have followed primatologists such as Richard Wrangham in looking in particular at the differences between ape and human diet. The former consists of fruits, leaves and other plant food. The latter adds to this animal foods in ever-increasing quantities, far beyond the quantities of meat eaten by chimpanzees. In contrast with the small sizes of ape territories that we have already mentioned, a carnivore-size home range for an ancestor like *Homo erectus* can for a community size of 100 mean a home range of 500 sq. km (200 sq. miles), away from the ecologically rich tropics. Big brains and larger communities therefore have major implications for diet and for the way the land was used.

So what changed in two-and-a-half million years?

The answer to this question is simple. Humans evolved big brains while other primates stuck to the ancestral ratio between brain and body size. Known as the encephalization quotient, EQ for short, this reflects the extraordinary growth of hominin and human brains. The end result is a human primate with a brain three times the size expected. We also know from the archaeology that technology became more complex. John Gowlett has charted this development and compared it to the trend in brain size (see Figure 3.4). He does this by examining the conceptual traits apparent in the changing stone-, bone- and fire-related technologies since the first appearance of stone tools 2.6 million years ago.

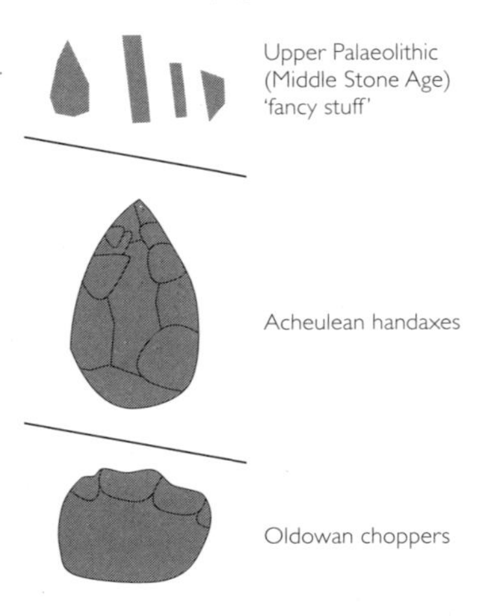

Figure 3.3: *Stone tools have traditionally been arranged in a simple ladder: choppers followed by handaxes, and then the 'fancier' Upper Palaeolithic toolkits – far from the full story.*

So far so good. But we already knew that brains and tools changed during human evolution. What the

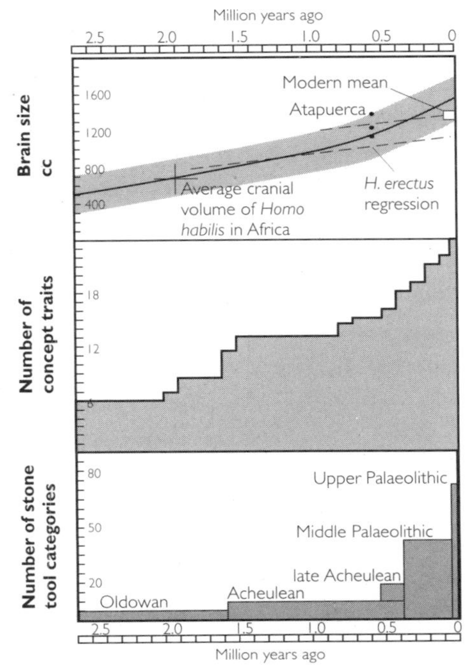

Figure 3.4: *Some major lines of evidence in human evolution brought together in a common timescale. Top: the well known rise in brain size. Centre: the introduction of new concepts in material culture, such as flaking stone, making bone tools, and using fire. Bottom: the number of stone tool categories in successive phases.*

social brain allows us to do is put the emphasis back on those individuals building their social lives with the essential resources of materials and the body's senses and chemical reward systems. Underpinning our understanding of these deep-history societies is the knowledge that the communities were up-front, personal and small scale. Here we examine the implications of increasing community size and then consider what happened when people started to live apart.

The size of communities and networks

The fact that community size correlates closely with neocortex size in primates and modern humans immediately gives us a handle on estimating likely community sizes for our fossil ancestors. The logic is very simple. If community size in monkeys and apes correlates with brain size, and modern humans sit exactly where they should do on the same regression line, then all fossil hominin species must lie between these two points – unless, of course, our fossil ancestors behaved in such a completely different way that they do not fit the general pattern for *all* other primates, including modern humans. That is so implausible it barely deserves serious consideration. Even so some people have, rather perversely, claimed that we cannot make any such assumptions about fossil species. The simple fact is: if brain size accurately predicts the size of social communities in monkeys, apes and humans, then it *must* do so for all extinct species as well – unless we are to assume that fossil hominins suddenly veered off-piste and did something completely un-primatelike, and then rapidly came back online just in time for the appearance of modern humans.

Nonetheless, some caution is not entirely without foundation. This is because the social brain equation is based on the size of the neocortex rather the whole brain, and perhaps better still on the size of the frontal elements of the neocortex (especially the frontal lobes). But we only have cranial volumes (in effect, total brain size) for fossil species, because brains themselves rot away very quickly after death and do not get preserved. Only the brain case is fossilized. Now, broadly speaking, the neocortex is a roughly constant percentage of the total brain across the monkeys and apes, so in the grand scheme of things this shouldn't matter too much, even if it might give poor predictions in the case of some individual species. Indeed, across primates as a whole, total brain volume makes much the same predictions about community size as neocortex volume does. However, there are individual exceptions. The gorilla and the orang utan both happen to have a very large cerebellum (the small brain-look-alike bundle that sits at the back of the brain, under the main brain hemispheres). The cerebellum manages neural messaging and coordination between different parts of the brain, and one of the important tasks it is involved in is coordination of the different parts of the body. Gorillas and orang utans are both very large, and manoeuvring their huge bodies in trees is a complicated business. It is no wonder they have unusually large cerebella. As a result, the cerebellum occupies a particularly high proportion of total brain volume, and their neocortices are, by comparison with other monkeys and apes, relatively small. Their respective community sizes are poorly predicted by total brain size, but are well predicted by actual neocortex volume. So we do have to be careful in making predictions about individual species, particularly the Neanderthals, who we will examine in detail in Chapter 6.

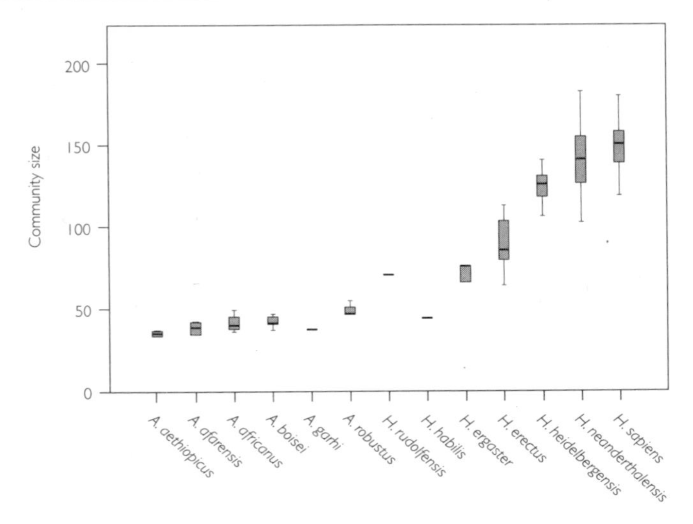

Figure 3.5: *Median community sizes (with 50 per cent range of certainty as box and 95 per cent range as whisker) for the main hominin species.*

With this caution in mind, we can use the basic relationship between brain volume and community size to estimate the latter for our various ancestors. The adjustments we need to make are for the effect of latitude if a species lives outside the tropics, since these species tend to be larger, and also for body size, since a bigger body will need a bigger brain to run it. Figure 3.5 plots the range of community sizes for all our ancestor species and their relatives, back to the earliest known hominins like Lucy and her kind. These results suggest that the earliest hominins, the australopithecines, were really just jobbing apes. They fall pretty much in the same region as the chimpanzees, with an upper limit on average community size of about 50 individuals. Individual communities can, and indeed do, exceed this value among living chimpanzees (the largest chimpanzee communities can be as big as 80–100 individuals), but such large communities are the exception rather than the rule. Chimpanzee communities become increasingly fragmented and unstable as they get larger, and it is this instability that creates the average or typical size of around 50.

Living apart but staying in touch

None of these predictions about community size factors in environmental change. Human evolution took place during a long-term trend in global climate to colder and drier conditions. After 1 million years ago this periodically led to major ice sheets in the Northern Hemisphere and the exposure of huge areas of the continental shelf in Indonesia – the tropical palaeocontinent of Sunda. Moreover, one other thing that we know changed during those 2.6 million years is that hominins moved from the tropics into the higher latitudes of northern Europe and southern Siberia. This occurred long before our own species started to move out of Africa about 60,000 years ago.

Now, mix together long-term climate change with hominins settling northern latitudes and you have a potent cocktail of environmental factors that the community sizes predicted from the social brain could not ignore. The impact would have been most keenly felt in face-to-face contacts. When food resources were closely packed and predictable then daily interactions with other members in your community were possible. This is the pattern that Robin Dunbar found when he sat in the 1970s with Gelada baboons in Ethiopia for many long wet months. However, if the food supply becomes more widely distributed, varies between seasons and overall the biomass of edible animals and plants decreases, then hominins have a social, as well as dietary, problem. The response to less predictable food supplies is to break up the group and replace fusion with fission, as we discuss further in Chapter 4. Here we simply note that there came a point when small-scale fission and fusion – which can be seen in all primates – was no longer possible. To stay put hominins had to cope with long periods of separation and yet still maintain the close social bonds that were negotiated in face-to-face contact. Of course the other option, pulling out and going where the grass was greener, was always possible too. On many occasions this is how these tiny populations adjusted to changing food resources. Equally they just became locally extinct.

Figure 3.6: *Washoe was one of the first chimps to grow up in a human home and participate in language studies.*

How long those enforced periods of separation had to be before the bonds began to crumble is difficult to estimate. If the bonds relate to rearing then they

Figure 3.7: *Christian remembers his former owners and administers a friendly lion hug. Animals have the capacity to remember social partners – but do they think about them, as we do, when apart?*

can be resumed on re-acquaintance. When Allen and Beatrix Gardner, who raised Washoe, a female chimpanzee, as their own child and taught her sign language, were reunited with her after an absence of 20 years, she immediately signed their name. As recounted by Jane Goodall, this was something Washoe did not do with the other chimpanzees she lived with. In 1960s London two young men bought a lion cub named Christian from the Harrods department store, raising it in their apartment until he became too big and was returned to Africa. A few years later they went to see him. The passion of Christian's embrace is heartwarming indeed and has become a YouTube hit (http://goo.gl/owJP). Christian certainly remembered his former owners just as they did him. And it doesn't need humans to be involved to happen. In her wonderful study of wild African elephants, Cynthia Moss recounts what happens when two sub-groups of this highly social species meet again after a separation: '[They] will run together, rumbling, trumpeting, and screaming, raise their heads, click their tusks together, entwine their trunks, flap their ears, spin round and back into each other, urinate and defecate, and generally show great excitement. A greeting such as this will sometimes last for as long as ten minutes.' Moss believes that such emotionally charged greeting ceremonies are all about maintaining and reinforcing the bonds between family members: an example of amplifying the emotional part of the social core (see Figure 3.2) through an elaborate greeting ceremony charged with feeling.

The big question for the social brain is not that Washoe and Christian could remember their human carers and act accordingly after long absences, but rather did they think about them at all when they were no longer face-to-face? Of course we can't ask them and devising a way to test the idea would be tricky, akin to determining if chimpanzees have a theory of mind. But what we believe differentiates us from all other primates is our capacity to carry on a social life as if others are present when they are not. The social life of a chimpanzee takes place in front of its eyes and around its nose and ears. By contrast, for humans it can also take place without any immediate sensory inputs from another individual. Social life takes place in our minds, in our imaginations. We constantly think about others from our various networks. We have their photos on our phones, their presents on our shelves and their words in our memories. Those others are the little voice of conscience that tells us someone else is looking as we reach for another sweet or consider doing something that we know would hurt them if they knew. And yet they may not be in the room or even in the town, country or continent where we are having that thought. We carry our social lives with us. We do this because we have the mentalizing skills, described in Chapter 2, of acute social reasoning that give us a theory of mind. This allows us to achieve something no other primate has managed: to live apart yet stay in touch. When it happened in human deep-history is something we will look at in later chapters. But when it did, it released us from the need to be in the proximity of others to be social. This could not have been done, as we shall explore, without amplifying the materials part of the social core – turning the stuff of nature into the things of culture. What such stretching of social relationships across time and space allowed was living at low densities with infrequent contact. The environment was no longer a brake on where our ancestors lived and the world lay before a hominin with such ability.

So why did social life change?

At the heart of the matter is the idea that our social lives drove the enlargement of our brains. We can show that brains got bigger. We can also see that there is a weak correlation with the appearance of new concepts inferred from archaeological evidence (see Figure 3.4). But what accounts for these trends and how can we marshal the evidence to test between theories?

Large brains come at a cost, because neural matter is very expensive to grow and maintain. This mainly has to do with the costs of keeping the neurons in a state of readiness to fire – a process that involves clearing the debris of neuronal activity out of the system, producing new neurotransmitters to allow the neuron to fire again, and maintaining the electrical charge across the neural membrane through the so-called 'sodium pump'.* The cost of maintaining neuronal readiness is roughly ten times that required for muscles to work, and the costs of actually firing the neurons (i.e. using your brain!) is considerably higher. Such expense also means that brains work best when supplied with high-quality foods. Moving from a chimpanzee diet consisting mostly of leaves, shoots, fruits and the occasional nut and monkey, to a hominin diet with increasing amounts of animal protein represents one such shift in diet quality; and as our brains grew in size so our carnivorous tastes developed. Larger brains also grew at the expense of another expensive tissue, the gut. As Leslie Aiello and Peter Wheeler have elegantly demonstrated, there had to be a trade-off to maintain equilibrium in the body's basal metabolic rate, the daily expenditure of energy measured when resting. Stomachs shrank as brains expanded. This had the consequence of reducing the processing power of the gut. One solution, as primatologist Richard Wrangham has explained in his book *Catching Fire: How Cooking Made Us Human* (2010), is to build an external stomach by cooking meat in a fire. This breaks down the enzymes making the food easier to digest. The evidence for when this happened is examined in Chapter 5.

All of these considerations surrounding the high cost of neural matter mean there has to be a very good reason for a species to increase its amount of brain tissue. But what has been the big benefit that increasingly large brains gave the various hominin species, including modern humans?

* The pump keeps the sodium ions on one side of the membrane, thus maintaining a difference in electrical charge between the outside and the inside. When the neuron fires, 'gates' in the membrane open and the sodium ions flow through, causing a change in the neuron's electrical charge. This charge flows down the neuron as the gates successively open, causing an electrical current to flow down towards the tip, where it connects with – and activates – the next neuron.

The advantages of larger brains and bigger communities

We see two main advantages for increasing the size of the community you live in, advantages that outweigh the massive costs of evolving bigger brains to cope with the greater cognitive load of remembering and acting on social information about others. These are security and reassurance.

Protection from predators and defence against others

The benefit of living in communities of increasing size is clear for primates as a whole. It is *protection* from predators. Primates, like baboons, that live in predator-risky habitats (terrestrial species, and species that live in more open habitats) have larger communities than those that live in less risky habitats. But this cannot provide the explanation for the evolution of large communities in any of our ancestors. It is not that apes' larger body size and formidable upper body strength (their arms are three to five times stronger than ours, no doubt to allow them to shin up trees in ways we simply cannot do) makes them immune to predation. Models of ape biogeography developed by Julia Lehmann as part of this project reveal that chimpanzees do not live where both lions and leopards occur: they can cope with one or the other of these formidable predators, but not with both. Indeed, even apes as large as gorillas are occasionally taken by leopards if caught unawares. The African apes' large body size certainly reduces their risk of being taken by a predator, but it does not eliminate it entirely – mainly, of course, because they also have young ones to worry about as well as themselves. Rather, the point is that, for humans and chimpanzees alike, the predation issue is dealt with at the level of the foraging group and not the community.

In chimpanzees, the grouping level that deals with predation risk is the foraging group (typically of 3–5 animals) and in humans it is either the night camp (of 30–50) or the foraging group (of 5–15 individuals). Humans are more vulnerable to predators, especially at night, for two reasons. One is that they live in more open habitats than other great apes (wooded savannahs as opposed to full-blown forest) and so have fewer trees to escape into when attacked; the second is that they are simply not as good as apes at climbing into trees in a hurry. As a result, humans need larger groups for protection than the other apes. Even so, we are not talking about the full community of 150, since this is rarely seen altogether at the same time and place (except for very rare, albeit important, ceremonial events).

So why do humans – and presumably all our ancestor species – need the larger, community-level of organization? There would seem to be two likely possibilities that recognize the importance of communities in *defending* against threats. The first possibility is to defend some kind of territorial resource. This has been suggested for chimpanzees, whose males seem to defend a territory that gives them exclusive access to the smaller ranges occupied by individual females and their offspring. It is a form of 'area-based polygamy'. Such systems are common among primates, where individual males defend territories that allow them to monopolize either groups of females or a set of individual females who each live in their own small territory. The only difference is that, in most of these cases, a single male defends a territory, whereas in chimpanzees groups of males (usually brothers) gang together to defend the territory, and share the females it encloses among themselves. It is possible that some modern hunter-gatherer communities are of this type too, but two factors suggest this is not so. One is that hunter-gatherer territories are rarely so assiduously defended as those of other monkeys and apes, probably because they are too large to defend. The other is that when territories are defended, they seem more often to be related to resources – and, in particular, what are known in the ecological literature as 'keystone resources', those on which you must depend when all else fails. For many hunter-gatherers, this will often be permanent water sources, as is the case in the Kalahari and Australian deserts. Nonetheless, given that the earliest hominins emerged from a great ape background, a plausible assumption would be that the earliest phase of australopithecine sociality was not so very different from that characteristic of African great apes: groups of related males defended communal territories to monopolize reproductive females.

A second possible reason why hominins might need large communities is that the conventional predators that regulate monkey and ape sociality had been replaced by another, altogether more formidable, predator – other humans. This is plausible, in that chimpanzees themselves show dramatic evidence of the dangers that can be incurred from one's own kind. Groups of chimpanzee males periodically patrol the borders of their territory, and deliberately murder any stranger males that they find. Although raids into other communities' territories occur, these patrols seem mainly to focus on locating target males rather than the acquisition of females or other economic resources in the way that they do among human societies.

In any case, the claim that this mechanism explains the need for community-level organization throughout hominin history implies that hominin population densities were very high throughout most of our evolutionary history, since raiding is invariably a response to limited resources and populations close to carrying capacity. While it is difficult to know what population densities were like in Palaeolithic times, it seems unlikely that hominin populations occupying woodland or open savannahs were ever at very high densities, not least because hunter-gatherer populations in such habitats today are obliged to live at low densities in very large territories. Thus, as an explanation for the early phases of social evolution during the early *Homo* phase, this seems at best unlikely. A better explanation lies in the keystone resources. These will include access to high-quality foods to support pregnant females and sustain infants who now, thanks to larger brains, are born ever more helpless and need nurturing for much longer.

Premium meals: cooperation and insurance

Higher-quality foods also have a habit of running away. The seasonal movements of animals can be fickle and they are often dangerous to capture. But once secured they can present the successful hunters with a bounty for sharing. And by sharing such premium food they can amplify the strength of the social bonds among community members. Richard Wrangham pointed to the importance of what he called growth, or premium, foods for getting primates to *cooperate*. These are the resources that females desire, because they give their offspring the best chances of surviving to reproductive age.

Archaeologists have followed through on these insights and looked at the array of food resources, both plant and animal, that was available and used by Palaeolithic communities. Borrowing heavily from biological studies, they have elaborated a set of principles grouped under the general heading of optimal foraging theory. In such research resources are ranked according to attributes such as size, weight and density in the landscape. Underlying this is the notion that the energy from food, measured as calories, is like a currency that can be managed according to rational economic principles such as costs being dependent on supply and demand. A matrix is then constructed to see what would be the best options at different times of the year, where in the landscape the people should be, how many of them and for how long they should stay there. This has not only transformed the study of animal and plant remains from archaeological sites, but also changed our understanding of what hunting and gathering meant. Rather than being a random, catch-as-catch-can existence, it became a rational game seeking solutions to the challenges thrown up by the environment.

This was good for the status of hunters and gatherers and countered Gordon Childe's gloomy assessment about their abilities. But it had a downside of suggesting that everything could be reduced to the choices centred on food. Food was certainly of vital importance and obtaining it efficiently and securely must have dominated much of their lives. The archaeologist Rhys Jones, who lived with Aborigines in Northern Australia, once said to us that these hunters and gatherers were always 24 hours away from hunger.

But for us the keys to understanding food lie in the implications for social cooperation. This takes two forms. First are the tactical demands of getting working parties together to either hunt or gather safely and with greater chances of success. This covers defence against predators as well as obtaining those foods that were needed to fuel expensive brains. Second is the strategic matter of planning for bad times. This is achieved by looking for help beyond your immediate community. Instead of restricting access to resources by defending them against all-comers, it is better to allow other people in. By linking individuals and their communities over very large geographical areas a form of ecological *insurance* is produced. Archaeologists refer to this as social storage: tokens exchanged for food in bad times, and vice versa in good. In other words, if conditions deteriorate where you happen to be ranging, then we will allow your community to come over to our range and use our resources for a while. Later, the reverse will be the case, and you can pay us back. Such a system works well, but it requires that the community has a territory large enough to cover a wide range of habitats. It won't work so well if community territories are small and consist of essentially the same kind of habitat. The more nomadic style that seems to have emerged with early *Homo*'s striding-adapted body form a little after 2 million years ago seems to provide a plausible basis for this (see p. 99). A community of 80–100 *H. ergaster* would, according to calculations made by project member Matt Grove, have covered an area of 130 sq. km (50 sq. miles) in tropical Africa, and that might have been sufficient to ensure that there was always some part of the range that had the required resources. But as those climates changed and greater seasonality faced those hominins who expanded northwards, so the driver to insure and link up with others became dominant.

Summary

What we have been talking about throughout this chapter has been inspired by the ideas underpinning the social brain. This perspective has allowed us to think differently about what is meant by social life in deep-history. It has thrown into focus the sorts of explanations that we can examine through archaeological evidence and we will shortly be turning to these. In particular we now have some idea of the scales at which we need to work and the importance of materials and senses to ancient individuals as they went about their social lives. We have even started to ask what sort of pressures – predation, defence, cooperation and insurance – drove the process. But what we have learned as archaeologists is the importance of thinking about the relations between people and the social core of materials and senses. Rather than concentrating on what may be rational explanations of why they hunted bison rather than reindeer or chose not to eat fish, as the isolated Tasmanians famously did 6000 years ago, we need to shift the perspective and see the role of food and other materials in creating relationships rather than simply meeting calorific goals. Archaeological explanations for the human story need to be relational (being social) as well as rational (being economic). Social life is not based on calories alone but on the relationships that emerge when things are made, exchanged, used and kept. And it is to these we now turn.

4

Ancestors with small brains

Cutting the chronological cake

In the social brain we have a powerful hypothesis: that it was our social lives that drove the expansion of our brains. But when do we begin to apply it to human evolution? And how do we test it? Some hypotheses can be tested by a single experiment: that objects of different weights fall at the same rate, for example. Our problem with the social brain is that its core idea is laid out in a graph that runs through many millions of years depending on where we decide to start the story. The factors that govern the shape of the graph are multifarious and interact in ways we do not always understand. As the archaeologist Ian Hodder once noted, when we have distributions with very few data points, such as skulls in the Pleistocene, usually more than one model can be fitted to them. And the scantier the data, the harder it is to test between the alternatives. There is no straightforward yes or no answer: this is the fascination and frustration of research into our deepest history.

Arguably the social brain hypothesis is relevant from the start of the story. The great apes have relatively larger brains than monkeys, and they lead complicated social lives as Frans de Waal so nicely illustrates in his book *Chimpanzee Politics* (1982). Here he portrays the stratagems of the ape-actors in terms that Machiavelli used to counsel his Prince. In that light, all today's great apes and their ancestors have been complex social beings for at least 20 million years. And it was from this heritage that the earliest hominins came.

These vast timespans are sparsely signposted with fossil remains (see Figure 4.1). Furthermore there is no hard archaeological evidence prior to the stone tools from Gona, in Ethiopia, 2.6 million years ago. But we can still use the social brain graph to cut the chronological cake into digestible slices. We wield the knife according to brain size, something that can be estimated with reasonable accuracy for fossil skulls. What we propose are three simple divisions:

- Living apes and fossil hominins with brains less than 400 cc.
- Small-brained fossil hominins with brains sized between 400 and 900 cc.
- Large-brained hominins and all living people with brains greater than 900 cc.

In this chapter we will be looking at the first two divisions, returning in Chapter 6 to the large-brained hominins and humans. The estimates of community sizes for these boundaries, shown in Table 4.1, are derived from the social brain equations described in Chapter 2.

The estimates for the time spent fingertip grooming those all-important social bonds are also shown in Table 4.1. These immediately flag up one consequence for a hominin's daily timetable of larger social communities with more interaction partners: an impossibly large proportion of the day has to be spent grooming (see box on p. 45).

Brain size	Estimate of community size	Proportion of daylight hours needed for grooming social bonds
Less than 400 cc (*Ardipithecus*, chimpanzees)	30–50	8–12 %
400–900 cc (australopithecines – graciles and robusts)	60–100	13–30 %
Greater than 900 cc (early and later *Homo*)	100–150	30–40 %

Table 4.1: *Hominins grouped by brain and community size and the amount of time predicted for social interaction.*

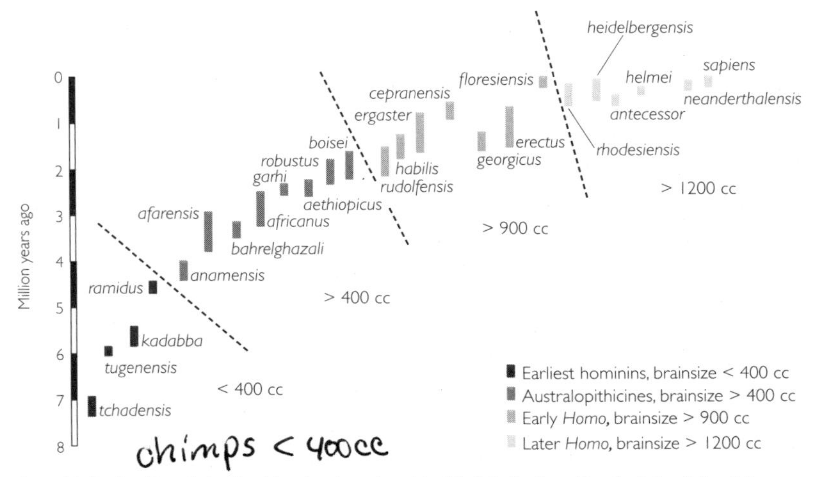

Figure 4.1: *Small- and large-brained hominins plotted onto the variety of fossil species.* Homo floresiensis *from* Indonesia *is a small-brained but late fossil that contradicts the general trend. The most likely reason for its diminutive size is the biological process known as island dwarfing, a reaction to the absence of predators.*

For the purposes of human evolution we have 7 million years to play with. This timespan begins with the molecular estimate for the timing of the split between hominin and chimpanzee ancestors. For the first 5 million years our ancestors were essentially apes who began to walk upright. They had ape-sized brains and we can summarize their history as 'ape to *Australopithecus*'. For the last 2 million years, bigger brains are strikingly evident – we summarize this history as 'from hominin to human' (see Chapter 5). In this second phase of human deep-history the role of the social brain is central to our account, because brains were rapidly increasing in size (see Chapter 6). In the earlier deep-history it is less obvious because growth rates were smaller. However, we are looking for foundations and we will argue that the two histories combine to form a continuum of socially driven changes.

From ape to *Australopithecus*: brains less than 400 cc

This period of our evolution covers the first two steps in Table 1.2. Fossil remains tell us that apes were distributed far and wide through the Miocene period (from about 23 to 5.3 million years ago), when temperatures were warm and forests extended across much of the Old World. They are found in Southwest Africa, East Africa, Greece, Italy, Pakistan and across to Indonesia. At some point as continents shifted and temperatures dropped, environments began to change, sometimes on a massive scale. Africa bumped into Asia, and as land masses reshaped their margins, three times the Mediterranean dried up, in the Messinian phase around 7 million years ago. Animals moved back and forth, and ancestral horses percolated into Africa, along with families of grasses which would feed them. Somewhere in this disturbed milieu the hominins began to evolve.

To track this evolution palaeoanthropologists have had to work backwards. When we first began our careers, there were just 2 million years of established record – beyond that, events were misty. Only a few finds charted the earlier long existence of apes, such as fossils from Rusinga Island in Kenya. Estimates of our ancestors' departure from apehood ranged between 15 and 5 million years ago. The current much fuller story has been shaped both by genetic evidence and numerous new finds. The foundations of later social change lie deep in this record.

The last common ancestor (LCA) of humans and chimpanzees existed some 7 million years ago. From that point hominins became separate from their ape cousins, probably because they were geographically isolated in drier and more seasonal environments. The earliest traces are found in eastern and central Africa. The oldest of all are of a creature named *Sahelanthropus,* found in Chad. *Sahelanthropus tchadensis* undermines the idea that something special about East Africa's Rift Valley prompted hominin evolution, because it lived thousands of kilometres to the west, probably close to a large lake that has since dried up. *Sahelanthropus* is known only from its skull, but that manages to be very informative. It is very ape-like, with a head rather like a chimpanzee, but already it shows crucial features that mark out the path of the hominins. In the skull the foramen magnum, the large aperture that connects nerves to the backbone, has already rotated from the ape position, indicating a habitual more upright position. The teeth also show changes, with the beginnings of a reduction of the canines, which is a hallmark of the later hominins.

Not everyone accepts *Sahelanthropus* as a true hominin, but traces of a similar creature have been found in the Tugen Hills of Kenya, with the name *Orrorin tugenensis*. Its finders, Martin Pickford and Brigitte Sénut, see *Orrorin* as the ancestor of all later humans. The skull is not preserved, but the femur or upper leg bone gives a good indication that *Orrorin* walked upright.

A far fuller picture of hominins several million years ago comes from the finds of *Ardipithecus ramidus*, which we shorten colloquially to 'Ardi', in Ethiopia. A joint American–Ethiopian team has found large numbers of finds over a 20-year period. Together these show us most parts of the skeleton, including the skull, a phenomenal achievement of recovery. Among these are the near-complete remains of an adult female who stood 1.2 m (almost 4 ft) tall and had a brain of 325 cc. Ardi lived 4.4 million years ago. One of her discoverers, Tim White, believes that *Ardipithecus* may be the 'stem hominin', and that *Sahelanthropus* and *Orrorin* are either very closely related to it, or actually in the same lineage.

Numerous mammal fossils give an idea of the environment in which *Ardipithecus* lived. To the initial surprise of many researchers, it was neither tropical forest nor savannah, but dense woodland with open patches here and there. Dietary evidence suggests that *Ardipithecus* was an omnivore, not eating particularly hard foods, but equally not yet the seeds, roots and animal resources of savannah grasslands – another significant observation, because for many years it was thought to be these that triggered the distinctive pathway of hominin evolution. Rather, palaeoanthropologists who conduct studies of tooth wear and isotopes in the teeth have shown that it was only after Ardi, and beginning 3 to 2 million years ago, that feeding on plant resources and animals from grasslands increased strongly.

Ardi: a small brain and a small social community

We begin therefore with a conundrum – how could the earliest hominins have been hominins when they still had small ape-sized brains? Our full answer we give at the end of this section, but the immediate answer for a fossil such as Ardi may be because they had no choice; it is very hard to evolve a large and expensive brain, and it could be done only under strong and prolonged selective pressures. Ardi's small brain size allows us to estimate that her social community had roughly 50 others in it, about the same number as in chimpanzees, whose brains are slightly larger but still under our 400 cc threshold. Fingertip grooming and near-daily contact among members seems entirely reasonable as the basis for Ardi's social life. And all of this could be accomplished in less than 15 per cent of the daylight hours, leaving plenty of time for finding food, eating, travelling and climbing trees to escape from predators such as leopards.

What else does Ardi tell us that might be helpful for starting off our social brains? The first point is that Ardi is very ape-like but not at all chimpanzee-like. If Ardi really is a stem ancestor, it shows that the modern African apes have gone down their own pathways in many respects and calls for a re-evaluation of the idea that chimps are the best model for the LCA. White and his colleagues believe that Ardi is far closer to the ancestral condition. Enough is preserved to show that these hominins were already upright walkers, or bipedal, at least for much of the time. This time the evidence is diverse and comes from various parts of the body. The pelvis has shortened remarkably compared with apes, and become more basin-like; the feet are far more foot-like than hand-like, but the divergent big toe was still able to grasp in climbing. The arms were long, but the thumb already opposable. The head lacks the great shearing incisors and canines that have evolved in the apes, and a significant thickening of enamel in the molar teeth indicates more emphasis on chewing.

Together these features point towards a new adaptation, but not one that required a larger brain – yet. In some ways, however, these earliest hominins seem to have evolved a new social platform that begins to give our starting point. These problems of early hominization were surveyed by the American palaeoanthropologist Owen Lovejoy long before he became involved with describing *Ardipithecus*. He and his colleagues set a lot of store on the early appearance of bipedalism and tooth changes in the hominins. They argued that a new social complex began on the ground, one in which males cooperated more, and in which they began to bring food to females. For apes, as Lovejoy pointed out in a classic paper published in 1981, there is a reproductive trap – each infant is highly dependent on the mother, with an inter-birth interval approaching several years, as much as eight years in the case of the orang utan. With becoming bipedal came a new freedom of social fission and fusion, with a succession of children weaned early and passed on to the care of others in sibling groups, or perhaps under the care of other adults, even including grandparents. Lovejoy and his colleagues argue that the finds of *Ardipithecus* strongly suggest three major changes in social structure – the hiding of ovulation in females, pair-bonding and food-carrying.

Unfortunately, neither anatomical nor archaeological evidence attest to the first of these, although at some point in hominin evolution it became indisputably important. Very probably concealed ovulation was there by the 'end of the beginning' – a point in the evolutionary story for which Ardi is currently our best representative.

However, the claim of pair-bonding is not supported by evidence from finger ratios as described in the box on pp. 66–68. Here members of the Lucy project showed that Ardi was very strongly polygamous. The social structure of gorillas might provide a model: a super-large silverback overseeing a harem of much smaller females and their offspring. A similar state in *Australopithecus afarensis* is suggested by the difference in size between that hominin's large males and small females (such as Lucy herself).

The third element of food-carrying seems plausible. The changes in locomotion, upper body form, teeth and increased sexual dimorphism (the difference in size between males and females) are probably the consequences of a new life on the ground that stimulated several interrelated factors of social change. One of these was that young offspring were highly vulnerable and needed more care. Meeting the demands of mothering and foraging at the same time was a huge challenge for females, so that there could easily be reproductive benefits in having assistance from an older generation, as the anthropologists Jim O'Connell, Kirsten Hawkes and Leslie Aiello have all argued. They drew particular attention to the role of females who were past reproductive age. A narrowly Darwinian analysis would be severe on these 'oldies' – if they can't reproduce then they are of no use to the group. But they do have a vital role to play as grandmothers. Their presence as childcarers increases the reproductive fitness of their daughters, since it is through greater investment and protection of children that more survive. The longer periods of infant dependency that were a consequence for our ancestors of bigger-brained children put childcare at a premium. Mothers, for example, could now search for food, leaving their children with granny. Rather than being tolerated, grandmothers were now celebrated. But encouraging longer lives also meant larger community sizes, as the three-generation society became the norm. In turn this could mean more people trying to live in the same place or maintain links because of their familial bonds, leading to greater pressures for fission and fusion of the community.

Ardi's place in human evolution

Let us stand back now and ask what these early ancestors can offer for our idea of a social brain. They were not human, but they are the first part of our story. They are on a trajectory that leads to humanity, but the exact route is unclear. Consequently we have to see them first as themselves; as having made a major adaptive shift from other apes. What drove this departure? In the first instance, it makes sense to see these creatures as 'dry-country apes' – apes who adapted to demanding environments in which regular rainfall and supplies of fruit and herbs could not be assured. The great apes of today all live in rainforests, and can live mainly from fruits and soft herbs, and their pattern of living depends on these being available throughout the year. With some variation it is a pattern common to chimpanzees, gorillas and orang utans, and so likely to be a very long-standing one going back many millions of years. But at the edges, it can fall apart easily – there are times when apes can scarcely get enough food. Those in drier habitats would have had no chance of finding traditional ape food resources through dry seasons, and – as the primatologist Richard Wrangham has argued cogently – would have needed to turn to new food resources. The 'extras' of chimpanzee diet already give us some hint of the possibilities – honey, insects and meat. Roots and tubers were another obvious option to replace the carbohydrates of fruit, but they would call for new adaptations, such as tools for digging, and were harder to digest. The new resources were distributed differently, probably more patchily, and at different heights above the ground, as well as under it. Bipedalism may be at least in part the best compromise for moving around and exploiting them. Although we may think of these things one by one, they add up to a new, complex life. The first differences between hominins such as Ardi and apes show sharply in their locomotion and teeth because these are the chief ways in which they engaged with the world for their subsistence. But minds and brains were probably also subtly challenged in the harsher environments, especially by the scarcity of food in prolonged dry seasons.

Work by members of the Lucy project shows just how sensitive most animals are to minor changes of temperature, rainfall or resources (see Chapter 2). This 'socioecology' can be plotted in equations and maps, and it also demonstrates the force of natural selection and the way it can demand rapid change as the price of survival. Over the course of the last 7 million years hominins changed much more rapidly than apes, and from this we can conclude that they did survive much greater selection pressures.

The teeth provide a strong hint that the changes did not just occur character by character, but as behavioural packages in which there were trade-offs. After Ardi, all the hominins acquire massive chewing teeth, but the interesting point is that reduction of the front teeth had already happened. In apes past and present those great canines are massive pegs at the corners of the mouth, useful for ripping, slashing and for sexual displays; male apes have far larger canines than females, so we know they were not simply needed for subsistence. For the hominins to forego these potent displays suggests a major change. Both *Ardipithecus* and *Sahelanthropus* show us that the change came early. In *Ardipithecus* there is no longer a morphological difference in the teeth between the sexes. There can be dietary reasons for chewing to take place further back in the mouth and, moreover, balancing the head on a bipedal body will be improved by not having these massive projections to weigh it down. But to lose such offensive and defensive weapons as ape canines must indicate a major change in behaviour. The shrinking of the canine teeth is fundamental. Between species, it meant that male hominins could not operate primarily by displaying and using their teeth – they would have to confront their enemies and prey with other weapons. Within a species such as Ardi, it took away a major marker between individuals, between the sexes. Outside vampire movies, human beings simply do not use the eye-teeth, those look-at-me canines, like apes or carnivores.

One of the oldest ideas in human evolution is that the freeing of the hands as a result of bipedalism allowed them to be used for holding sticks and other weapons, as well as carrying food. If so, dental weaponry and its threat displays could be transferred to what was held in the hands. However, if *Ardipithecus* was – despite its bipedalism – still also a habitual climber, as the anatomy of the shoulder, hands and feet suggests, it probably did not do so 'staff-in-hand'. It would have climbed for foraging, or avoiding predators, but at some point there came a change of commitment in these small-brained apes, from tree to ground, and *Ardipithecus* must be very near to the cusp of this development. And that is why we call her a hominin, while recognizing the conundrum this produces because of her small, ape-size brain.

In essence we are saying that a shift of everyday tasks from teeth to hands centred on interaction between individuals and had profound social implications. Most importantly from the archaeologist's perspective, we believe that the technology was from the start essentially 'socially embedded' – a tool both of and for socializing.

Australopithecines: the jobbing apes with brains greater than 400 cc

The australopithecines were successful, of that we can be certain. With their larger brains, they spread out across Africa. Their success also shows in a new variety of species – a phenomenon known as an adaptive radiation. This was not as spectacular as the radiation of sea mammals, or of Old World monkeys, but numbers of new species emerged.

The scientists who have described *Ardipithecus* question the old position – that a chimpanzee-like LCA moved towards being human through the 'transition' of the australopithecines. Instead they argue that the australopithecines have to be seen as a distinctive adaptation in their own right. A problem for a smooth transition is that they succeed *Ardipithecus* so quickly: White and Lovejoy believe that the hind-limb evolved very rapidly indeed. An alternative would be that *Ardipithecus* is a sister-species to one of the australopithecines that became more fully bipedal. Either way, the australopithecines appear on the scene at least 4 million years ago with *Australopithecus anamensis*, a species known from northern Kenya.

Around 3 million years ago we find australopithecines from the very south of Africa up to Ethiopia, and to the west at least as far as Chad. In the simplest analysis we can see them still as upright apes, and we have very little evidence that they used hard tools. They resemble apes in a number of respects – their cheek teeth were massive, their arms were very long, and the thorax still cone-shaped, an indication, together with the long curved finger bones, that many australopithecines may still have spent part of their life in the trees. But there has been a quite crucial adaptive shift: the hind-limb has given up on grasping, and become fully adapted to walking on the ground. The most direct

Figure 4.2: *The fossil trackway at Laetoli, Tanzania. Dated to around 3.5 to 3.8 million years ago, it preserves the prints made by two adults and a juvenile as they walked unhurriedly across a plain covered by a thin layer of fresh volcanic ash from nearby Mount Sadiman.*

Figure 4.3: *Mary Leakey, one of the most prominent Stone Age archaeologists of the last generation, works on cleaning off sediments at Laetoli.*

evidence of this comes from the famous footprint trails at Laetoli in Tanzania. These show us a small group of hominins walking together, more than 3.5 million years ago.

From a social brain perspective the australopithecines are the first hominins to break the 400 cc threshold for brain size. Below that threshold are Ardi, *Sahelanthropus* and all the living great apes and monkeys. Gorillas do exceed the 400 cc but as explained in the previous chapter this is due in large measure to their big hind-brain, the cerebellum, an adaptation needed to coordinate moving their great bulk. Neocortex size is what matters for the social brain.

There are a number of 'gracile' australopithecine species – *Australopithecus anamensis*, *africanus, afarensis, garhi* and *sediba* at least. In addition there are robust lineages marked out by their massive jaws – *Australopithecus robustus*, *boisei* and *aethiopicus* (set by many scholars in a separate genus, *Paranthropus* – hence 'paranthropines'). Brain sizes are variable, but as leading palaeoanthropologist Bernard Wood has shown, they fall in a range from 450 to 570 cc. As a guideline, these translate into community sizes of 60 to 80 interaction partners. This in turn exerts an upwards pressure on the daylight timetable, with the larger communities needing to spend around 20 per cent of daylight hours on social activities such as fingertip grooming.

All our social questions arise again with the australopithecines. Apart from the Laetoli footprints, we have further evidence of social groups from fossil evidence at Hadar in Ethiopia. If canines were no longer a marker of sexual difference, then quite certainly there were still large differences in body size between males and females. Lucy, the famous type specimen of *Australopithecus afarensis*, was little more than a metre tall, but other finds from the same Hadar region of Ethiopia are of strongly

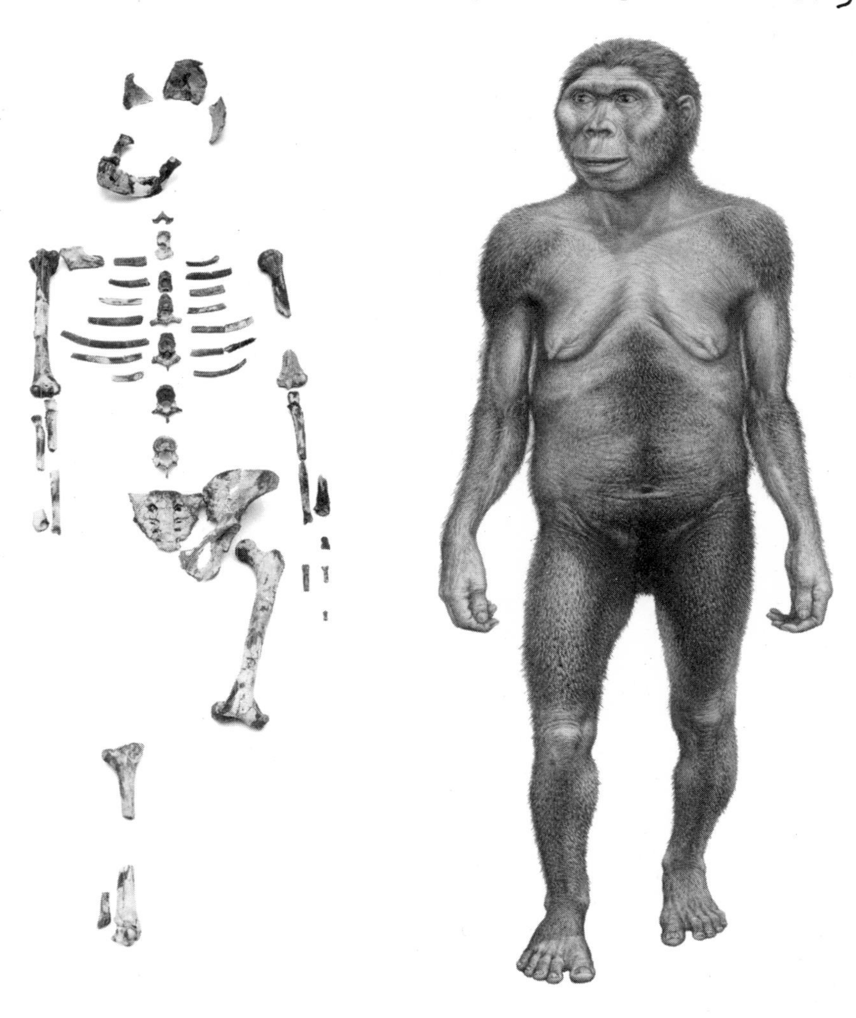

Figure 4.4: *Lucy is the best-known representative of* Australopithecus afarensis. *Skeletal remains of this species show that the brain was still small but upright walking had developed.*

built male specimens, probably as big as male chimpanzees. This is our best primary evidence that the australopithecines may have been closer to apes than to modern humans in their social behaviour. Comparative studies of primates indicate that societies combining large-bodied males and small-bodied females usually operate on a 'harem' model. Gorillas provide an example: the social group often consists of one senior male, the silverback, and a number of females, together with infants. As males become adult, they often have to move away from the dominant silverback and eventually they may succeed in building their own new social group. Living in communities must have been more than a luxury: if they were no longer arboreal, australopithecines must have been vulnerable to large ground-living predators – the big cats and giant hyenas especially. This provides an answer to the question, how could Lucy be so small? An obvious conclusion is that her community provided much of the protection. Indeed this must have been so, because bipedalism entailed having small infants running around. In living apes, infants are carried by the mother until they can begin to operate in the trees. Finds from South Africa suggest that the dangers were real – Swartkrans Cave preserves large numbers of australopithecine remains, but very few of later *Homo*, even though stone tools show a later human presence. Following the arguments of Bob Brain, the australopithecines at these sites were the hunted rather than the hunters, the regular prey of leopards and other big cats.

A most striking thing about the australopithecines is that from their original 'platform' they evolved into at least two separate adaptations. Upright walking creates a biomechanical issue for the head – it is up on top. It cannot easily and economically support both massive jaws *and* a large brain. Somewhere, somehow, in different local environments, natural selection appears to have favoured the one or the other. One group, the robust australopithecines (or paranthropines), developed massive jaws and teeth. The large muscles that worked them extended over most of the skull. There must have been a necessity for processing large quantities of low-grade plant food. Our nearest parallel is with the gorilla which, though a primate, can be described as a herbivore because it eats large quantities of stems, bark and pith.

The other direction was to favour brain. We do not see this tendency straight away, but some species at least did not evolve the specialization of the robust australopithecines. The gracile australopithecines all did without the massive musculature of the robusts. And if we are talking bigger brains then we are talking the earliest representatives of our genus, *Homo*. The transition from hominin to human is arbitrary – fossil skulls have dropped in and out of the human club as palaeoanthropologists change their minds in the light of new evidence and new concepts of what makes a skull human. All we are saying here is that the transition began among hominins with brains in the range of 400–900 cc.

Beyond jobbing apes: the earliest *Homo*

Thanks to the vagaries of preservation and discovery, some periods are represented by more finds than others. It happens that from 2.6 to 2.0 million years ago, we have fewer cranial remains than before or after. The beginnings of *Homo* probably lie exactly within this period, but the chief evidence for that are the varied remains that occur just after 2 million years ago. Two or three species had already appeared, suggesting deeper ancestral roots to our genus, of perhaps half a million years. *Homo* can be diagnosed by specific detailed features of the cranium and teeth, but the overriding indicator is a major increase in brain size. In some cases, such as the famous KNM-1470 specimen from East Turkana, this is 800 cc, nearly twice average ape size. Since its discovery this cranium has always been classified as *Homo*, but assigned variously to either the species *ergaster* or *rudolfensis*. In addition, some *Homo habilis* specimens have an average brain size of 650 cc, a rise of about 50 per cent above the australopithecines with whom they are sometimes lumped, but still well below *Homo rudolfensis* and the later *Homo ergaster*. The truth is there is no brain size that allows us to say that above that line are *Homo* and below it some older non-*Homo* ancestor. At the margins there will always be fuzziness, as the naming and renaming of fossil skulls tells us.

However, from a social brain perspective we can expect community sizes of between 80 and 90 for these earliest *Homo*, whatever they might eventually be called by palaeoanthropologists. Such numbers are well above the limits seen in modern apes and monkeys and above those predicted for the australopithecines. This means that about a quarter of their daytime activities must have been spent grooming to create, confirm and reinforce social bonds. Before long something had to change.

Why did this brain growth happen? We would argue that the preconditions had been met for a new greater sociality to take off. This is the social brain in its truest sense: social lives driving brain growth. One of these preconditions was the stone tool, which provides a prime example of the process of amplification, boosting the signal from an existing adaptation to meet the challenges of larger community sizes. How exactly this manifests itself in social brain terms will be explored below. First we must take a closer look at the role of tools.

The trail of tools

We are still fascinated by technology, and the extra momentum that it gives us in the world. Passengers on Concorde could look at a display at the front of the cabin and see when they reached twice the speed of sound – thus joining the select Mach 2 Club. Archaeologists could similarly have a '2 million years club' for those who have found and worked with the earliest stone tools. Some of us would never be quite sure if we were in the club or not, as the dating of events is not always that exact. But undoubtedly the 2-million-year tool barrier was breached in around 1970, with new research centred on Lake Turkana in northern Kenya. A further generation of research has pushed back the dates even further, to 2.6 million years at Gona in Ethiopia, and we can be fairly sure that the tools are telling us about a major step along the road of human evolution.

What is so special about stone tools? Archaeologists have been able to use them as markers of the past for more than 200 years now, and one great value is that they do leave an almost indelible record of past human presence. Having a record is infinitely better than having no record, but from there the hard part starts – we need to know 'who made them?', 'what with?', 'when?' – and 'were they actually important?'. These questions will keep us in close contact with our social brains.

To work out the answers we have to look first at a greater world, not of humans, but of animals, pursuing their lives and using an intriguing variety of means to help themselves along. Most of them, whether mammals, birds, reptiles or fish, use their own body as their entire means of physical operation. If it cannot be done with tooth, paw, beak or claw, it will not be done. But many animals do find a way of engaging with the world by using something extra, outside themselves (extra-somatic)

– they use some bit of external material to influence something else. And that's what tools are all about.

Some of the cases are specific and simple. A bird will drop a snail on a stone to smash its shell. Macaques will wash potatoes (water is the tool). A dolphin will put a sponge on its nose for stirring up the seabed. Other cases involve a good deal more complexity. The star performers – apart from New Caledonian crows – are capuchin monkeys, chimpanzees and, of course, hominins and modern humans. The picture is intriguing, because clearly it does not all depend on close evolutionary relationships. The great majority of monkey species do not use tools, not even the closest relatives of the capuchins.

Apes are by most measures more intelligent than monkeys and they are certainly more encephalized. But they do not all use their abilities for toolmaking. Indeed only the chimpanzee does so in a systematic way. Fundamentally, as we have outlined, their intelligence is directed towards meeting social and ecological challenges. Primatologist Richard Byrne has pointed out that activities like preparing plant foods and constructing nests also require some of the same abilities as toolmaking and use. This behaviour too is learnt in a social environment.

A curious aspect of our ancestral family tree and its phylogenetic branching is that we and the chimpanzee are more closely related to one another than either is to the next most common toolmaker, the orang utan. The orang utans certainly make tools, but not for all occasions. The gorillas scarcely use them at all; and our other closest relative, the bonobo (or pygmy chimpanzee), rarely does so. In this pattern we cannot make too much of the chimpanzee being our closest living relative, particularly as it shares this distinction with the bonobo, but we can use the chimp as a possible indicator of the kind of tool-using that our ancestors began with. Of course it is not sensible to use any present behaviour as a direct key to the past, but the chimps help us to see how a range of simple tools can be used for a variety of tasks. For this reason some see the chimpanzee as the best model that we have for the last common ancestor of hominins and apes. This view is championed by chimpanzee researchers such as Bill McGrew, Richard Wrangham and Andy Whiten. Others would argue that the chimps have acquired major new adaptations of their own. But it is still convenient to regard them as the best *living* models.

Figure 4.5: *A chimpanzee using a stone to crack nuts. Such behaviour varies between populations of wild chimpanzees and takes many years to perfect.*

One thing is certain, chimpanzees use tools more often and with more variety than any other animal apart from ourselves. All chimpanzee populations use tools, but not always the same ones, and although most of the tasks are for subsistence, many of them are not vital to life. The main part of chimpanzee subsistence comes from fruit and herbs, and tools are not essential for exploiting them. Then the tool-use divides into hard and soft – stones used for pounding, and softer stems and leaves for termite-fishing and mopping-up. Some chimpanzee tool-use greatly catches our imagination. At Fongoli in West Africa the chimps hunt bush babies with small wooden spears; elsewhere they use sets of tools to break into underground termite nests, or into beehives to get honey, defying all stinging and biting.

Amid the huge interest that ape tools have engendered we can draw some firm conclusions:

- Tools have come out of a social world.
- What they do does not relate to main subsistence activities.
- Technology says quite a lot about the 'extras' that are important in chimp life.
- 99 per cent of chimp tools would not survive so as to generate a clear archaeological record.

Do the tools say anything about intelligence (defined here quite narrowly as solving food-getting problems)? When it comes to exploring the significance of tools, primatologists and archaeologists share a difference. Primatologists cannot ask their subjects why they do and do not do things, because the apes do not have language. Archaeologists cannot ask their subjects, because they are dead and gone. In fairness, apes who do not make tools appear as intelligent at solving the challenges of social life as those who do. But intelligence, a slippery concept at the best of times, does seem to be a requirement. Somehow among chimps the cultural knowledge is passed on, because other chimps see and learn. That is what we mean by tools starting in the social world. The tools are moderately important to chimps, but it seems they would be far more so if they lived in even a slightly different environment, so that a higher proportion of their food could be extracted only with this help.

The social role of tools

Overall there are two ways that we can look at tools:

- The world is predominantly a social world, and tools are a kind of accidental by-product, an epiphenomenon.
- Materials from which tools are made are a core part of the scaffolding within which our social life emerged.

The strongly marked presences and absences of tools among groups of closely related species, in both birds and primates, support the first view – even in the case of modern humans. Some traditional Australian aborigine communities had immensely rich social lives, but a small toolkit. Equally the story of human evolution as told by archaeologists such as Childe stresses the importance of the second point. The technological base seems to have been essential for each major economic leap forward in human evolution. However, each economic leap is also tied to social developments.

We can reconcile these points to some degree, and that is important for relating technology to the social brain. First, it is plain that tools give us a very powerful index for looking at the past, and the same is true in the present. Even an ornament is a tool for projecting information about its wearer. Above all, tools are almost always made or used in a social context. This is just as true for primate tool use. In recent years primatologists have felt that they needed to prove the existence of 'social learning', but there is only one real alternative, that all tool use stems from new invention by individuals. That is hardly possible, and the essence of tradition is that individuals learn from others, through contacts that can only be regarded as social. These arguments allow us to see technology in human ancestors as fundamentally socially embedded. Without it there would have been no transmission of ideas and techniques, no inheritance of successful ways of coping with the world.

However, from the start, technology has effects resulting in social activities that can only take place because of previous technical action. For instance, we can only live in a hut because we have built it. Living in it we can have new social opportunities that create new technical opportunities. In such ways feedback loops tie technology deeper into our biology and sociality, that process where we amplify the signals for building social life by strengthening the bonds that bind. A branch of biological and anthropological science has recently arisen which describes this process as 'niche construction': human beings have made themselves a massive cognitive and technical niche, but its underpinnings are social and ideational. Humans can claim that we actually build the niche that we live in. Many other animals do this too, but no niche is so varied or complex as ours. We are one with our material worlds and the environments we have built to live in – but humans take this much further. As anthropologist Maurice Bloch put it, the key capacity of humans is to live in the imagination; or as we would put it, to explore the possibility of intentionality above the fourth order (see Chapter 5). The important point is that this is not something that started with the towns and cities of the Neolithic revolution. It began deep in our ancestry and the stone tools from Gona merely mark the currently known beginning.

By contributing to building our distinctive niche these ancient tool-using skills, learned in a social context, point to a different understanding of the organization of the hominin mind. We favour an extended mind, one that is as much chipped stones as it is grey matter, and a social cognition that is widely distributed in the environments that hominins live in. We don't mean literally that brain existed in the stones, but that hominin minds mapped and modelled all these ideas at a fundamental intuitive level. These ideas are explored in the box overleaf. They point to continuity between ourselves and the makers of the earliest stone tools – a far cry from earlier accounts of hunters and gatherers and Stone Age ancestors as more childlike and irrational in their behaviour. The latter was the firmly held opinion of one of archaeology's greatest exponents, and committed Darwinian, General Pitt-Rivers, writing in 1875. Such views about who had the power of rational thought, and hence the intellectual faculties to innovate and solve problems, can also be found in Childe's assessment of the savage stage of social evolution (see Chapter 3).

The hard record: the when, what and who of tools

Somewhere in this evolutionary matrix stone tools did actually appear. The oldest that we know of come, as we have noted, from Gona in Ethiopia, quite close to the *A. afarensis* finds of Hadar, but at higher levels. They are dated to about 2.6 million years ago, whereas the youngest *A. afarensis* finds are older than 3 million years. Discoveries in 2010 at Dikika, further south in Ethiopia, suggest that cutmarks may have been made on bones at around 3.3 million years ago, but the evidence is currently disputed. The idea of such an early date should not be a problem, considering the proficiency of other primates in using simple tools, but the absence of tools at some other well-known localities would be hard to explain. The extensive landscapes of Laetoli were explored by experienced palaeoanthropologists well-used to stone tools, but nothing was ever found. Equally, the older australopithecine caves in South Africa lack any signs of stone tools. When they appear, as if out of the blue, the tools have two huge uses for palaeoanthropology. First, they are markers of hominin presence, the nearest equivalent of modern smartphone technology that tells you where a person is, or

Types of mind

Hominin brains can be measured and examined. No doubt great leaps forward will come in the next decade from MRI and CT scanning to see which bits of our brain light up when we make different stone tools or use language. That will be an experimental revolution comparable to the impact that field studies of chimpanzees and baboons have had on our generation of researchers.

But what went on in the minds of these hominins is far trickier to determine. Minds don't preserve, only their physical products survive. However, the important first step is the model of the mind that is used, because which one is chosen then determines the sorts of questions raised and the way the archaeological data are approached.

The best-known model of a mind can be traced back to the 17th-century French philosopher-mathematician René Descartes. He characterized the mind as a rational instrument that solved problems outside the body. This was a major break with medieval ways of thinking about the world and laid the foundations for science and medicine as we know them today. Since Descartes there have been many elaborations on his juxtaposition of cognitive processes as internal:external and his other dualisms of mind:body and object:subject.

Many metaphors have been employed to discuss these inner workings and today, unsurprisingly, the computer is widely used. In his pioneering book *The Prehistory of the Mind* (1996), archaeologist Steven Mithen used the concepts of mental modules to structure his evolutionary account. These modules, he argued, were different intelligences and dealt with such spheres as social, natural history, language and technology. This compartmentalizing of the mind provided much-needed purchase on what was otherwise a woolly concept, cognitive archaeology. When it came to explaining change Mithen used the idea of cognitive fluidity that linked in ever-more complex ways the modules of the brain. In his scheme the natural intelligence module merged at some time in human evolution with the social intelligence module. Those reindeer, which were once simply food, now became good to think as well as good to eat. They were used as clan totems and do much more than simply supply calories. While we disagree with the modular mind – too rigidly rational for our taste – Mithen's striking evolutionary scenario is at heart a scheme that stresses the growth in the relational bonds to which our cognition is supremely well adapted.

In a similar way archaeologist Thomas Wynn and psychologist Frederick Coolidge have developed a model of Neanderthal minds. Rather than modules they stress the importance of memory, short- and long-term. In *How to Think Like a Neanderthal* (2011) they discuss a wide range of topics, Neanderthal jokes, dreams and personality and as a result challenge the preconceptions about this big-brained, brawny hominin.

What we see as the link between these two important books is their starting point, the Cartesian rational mind. This model is fundamental, but we have also noted the crucial emotional base for human behaviour (see Chapter 2), and we would also stress a more relational quality, the ability to make associations between people and things that is so prevalent in human cognition. As a result we would break down the barrier between minds 'in-there' and the world 'outside'. With a model of an extended mind we have to change our definition of what we class as a mind. It is no longer just the grey matter inside our skulls. It extends beyond our skin and includes the things we interact with and the environments that surround us. These can be of our own making as in niche construction. Our minds are as much the cups we drink from as the chair we sit on and the neurons firing in our brains when we perform these actions. Minds are not just about being rational in our approach to others but also relational in a truly social sense. Our social cognition is therefore not only to be found in the prefrontal and temporal cortex where memories and information about others is stored. It is also to be found in the accumulation of artifacts, their shape, touch, taste and smell. In that sense our social cognition is distributed throughout the worlds we live in, a basic part of the niche we have built.

Other animals have this type of distributed cognition. But we have the capacity to extend who we are and what we can do, our agency on the world, through the things we make, buy, exchange, keep, treasure and throw away. This makes for a very subtle intelligence that both reflects on and derives from our intense social life. The big question for the history of such a social brain is the extent to which some or all of our ancestors had the same ability to use the opportunities of an extended mind to pursue ever-more complex social arrangements. One of these was to break the shackles of face-to-face social lives. Unleashed from the immediate, hominins now carried on a social life when they were apart by thinking of others and regulating what they did by the thought of what others might think of them. Among the results, we can see that early hominins, by 2 million years ago at least, operated on far larger scales of time and space than their cousins the apes. Was that the start of the modern mind?

in this case was. Until this point we have needed hominins to die to be seen. Second, once present, the tools tell the story of many everyday activities, and how problems were solved in doing them.

To give us these benefits the tools do not need to be sophisticated. The earliest ones are simple stone flakes struck from stone cores, but they were made with confident consistency indicating skilful practice. Altogether the tools mark out a new domain for our studies, alongside the hominins themselves and their environments. So far, all the tools older than 2 million years have been found in Africa, just as all the australopithecines have been found in that continent. The tools are lumped together conventionally as one tradition, the Oldowan, named after Olduvai Gorge in Tanzania. Here Mary Leakey first described the tools from large sites near the base of the Gorge that date to about 1.8 million years. Later explorations revealed similar tools at a number of other places along the Rift Valley, from Ethiopia down to Malawi, and they are also known from several of the South African cave sites.

The patterns of use led archaeologist Glynn Isaac to call the toolmakers 'the first geologists'. Right from the oldest sites, they knew where to find the most suitable rocks. They used to travel to such locations, at first involving distances of 3–5 km (2–3 miles). Henry Bunn has traced these activities to the east of Lake Turkana; Rob Blumenschine and colleagues have done the same around the Olduvai lake. It stands out that there are often different patterns for different materials. The hominins were also highly selective, only transporting the rocks that had the best qualities for stone knapping. Archaeologists generally find sites that are single focuses of activity over some limited past period, perhaps days or weeks. But studies by Nick Toth and Kathy Schick have shown that they are always part of a wider network – every Oldowan site that they investigated showed signs of both imports and exports of stone. Very roughly we can suggest that the hominins did some first working on the site where the raw materials were to be found – this eliminated deadweight of material useless for tools; then they carried the useful material to a favourable location or occupation site where there they made finished tools, which were then taken away to be used.

All this gives us vital social information – about the distances hominins travelled and the scale of their operations. And crucially from the start, this was far bigger than the territory sizes of apes. Just as clearly it was not an individual exercise, since the quantities of rocks moved are sometimes staggering. This was a collective exercise of shared intent, and one that gives some insight into the size of the social communities. It is indisputable too that, like carrying larger brains, carrying rocks is an expensive activity that must have paid for itself. Tools must have opened up the way to new resources. Meat, tubers and roots were certainly among them. Stone tools also facilitated the making of further tools – such as sharper sticks for protection or digging.

There is just enough evidence to show that a range of tasks was undertaken. Hominins tended to prefer compact materials such as basalt for heavy tools, and fine-grained shiny materials like chert for sharp cutting edges. Some sites in Olduvai Bed I show this conjunction, which we find again at sites in Israel far later in time.

The when and what questions surrounding stone tools are relatively straightforward. Who made them is much more problematic. Most of the earliest stone tools are not associated with any fossil remains. When they are, as in the lowermost beds at Olduvai Gorge, the processes of deposition, sorting and movement due to natural causes such as floodwater have to be taken into account. Famously in Bed I Louis Leakey first found the skull of a robust australopithecine, then known as *Zinjanthropus boisei*, and only two years later a gracile hominin from the same level that was named *Homo habilis*. Zinj's brief window of fame as the maker of stone tools was eclipsed by the discovery of 'handy man', an eclipse that had less to do with the evidence than the assumption that *Homo* had to be the toolmaker: 'man the toolmaker' in palaeoanthropologist Kenneth Oakley's highly influential little book of that name (1949) that by 1972 had gone through no fewer than six editions but with the central message intact. But now *Homo habilis*, who probably *was* a tool-user, has for some palaeoanthropologists lost his status as a member of *Homo*, placed instead among the australopithecines.

The honest answer is that we just don't know. On balance, and in the light of all the non-human instances of toolmaking that have come to light since Oakley's book, we strongly suspect that toolmaking existed variably not only in small-brained hominin species such as *habilis*, *africanus* and *robustus*, but also perhaps to differing extents in contemporary populations living in different parts of Africa. What is clear is that the 2.6-million-year-old Gona stone tools pre-date by some margin the first glimpse we get of the earliest *Homo*. We return to this issue in Chapter 5 when we look at the innovations of fire and handaxes.

The advantages of tools

We can round up the early tools with two observations and a question:

- They solved problems.
- They transmitted ideas.
- Did they have design?

At first sight the question seems trivial, but it is a fundamental one. Of necessity any tool brings together concepts, and this idea of concepts is at the heart of all our communication and social networks. If a tool can be seen as a net of the features that make it up, it is a small *world* that embodies these – much as our network of contacts is a bigger net, a tool for social survival. Even a simple stone tool must have a working edge, with properties; it must have sufficient extension to be held; it must have an appropriate mass. Is that enough to be a 'design'? It will be easier to argue this case later on, when tools become more sophisticated, but in essence any tool is a small world within the big world, and our dealings with it have similar parallels. A tool allows us to pay attention to the detail, to study something that might have taken five or ten minutes to make but which represents hominin technology across half a million years. And this attention that we bring to the attention they applied to the job in hand makes this a supremely social technology. By this we mean something that takes its significance from commonplace acts of observation, social interaction and learning. And as we saw with niche construction, we cannot separate the tool from the hominin who made it. Oldowan stone tools were a part of their makers' distributed cognition (as explained in the box on pp. 106–07), as much a part of their mind as the neurons in their brains. For this reason anthropologist Leslie A. White talked of stone tools as 'symbolling' – anytime we see one, we engage not just with the object, but with its wider meanings. So, if we are correct in arguing that social life drove the expansion of our brains, we have to accept that technology, and the world that surrounded every hominin and through which they made their way, is also socially conditioned.

Technological change

When we try and weigh up the changes in technology, and the drivers of change, we can ask where the social brain fits in. A straightforward explanation for the technology, as used by previous generations, would be simply that it paid to be cleverer so that you could develop and use the technology and gain resources. If that required more brain, it would be favoured, and selection pressures would ensure that it evolved. We can now see that this technology-driven explanation is too simple. Most people, most of the time, are not actively using technology, and so it should not be demanding a great deal of brainpower. Modern populations vary from those who have huge amounts of technology, to those who have very little, but this seems to have no bearing on brainpower or intelligence. Even a chimpanzee, apart from making its own tools, can make very effective use of many human tools, often self-taught.

In contrast with a technical scenario, the social brain hypothesis would predict that in a hominin population any increase in brain size would be socially driven. A simple explanation for events around the emergence of *Homo* (as well as of technology) is that hominins were needing to live in more open landscapes, and that to do this they needed to operate in larger numbers, over longer distances, over more extended timescales – and that they needed larger brains to organize all this, not to mention the social ramifications of living apart but staying in touch. The stone tool transport distances give the clearest evidence of the scale of movements. Oldowan tools are rarely made from rocks brought from more than 5 km (3 miles) away, though even so they do appear to have been selected for their knapping quality. In the later Acheulean these distances treble, with several examples showing the transportation of stone over much greater distances. Once again the hominins who made handaxes show a preference for certain raw materials. In her study of obsidian Dora Moutsiou, a Lucy project research student, showed that the average distance travelled for this volcanic glass in East Africa in the later Acheulean was 45 km (30 miles) from source, while maximum distances of 100 km (60 miles) are known from Gadeb in Ethiopia even 1 million years ago. Without more evidence, we cannot talk yet of the questions of exchange that we will move onto below.

We can see these distances as definitively illustrating an expanded scale of group fission and fusion of the kind that colleagues such as Filippo

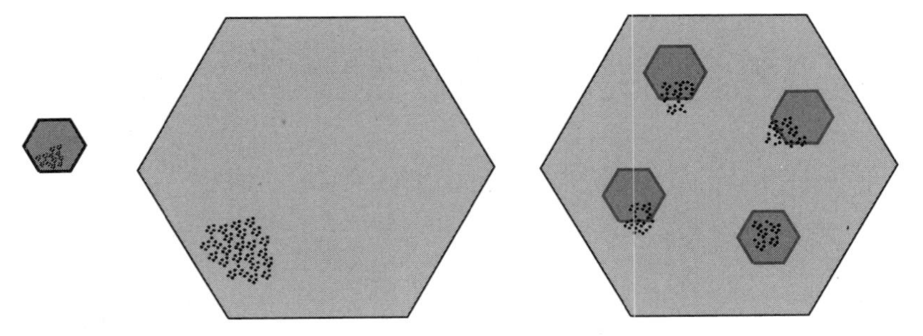

Figure 4.6: *Left: Chimpanzees generally operate in small territories and in general the community can congregate or keep in touch by vocal communication. Centre: In the far larger territories frequented by hominins, resources rarely allowed the community to come together in this way. Rather (right) the community tended to subdivide into bands, flexibly organized, and often centred on waterholes and areas of richer resources.*

Aureli have discussed. From their earliest appearances in East Africa, stone tools demonstrate systematic movements across several kilometres and they may be telling us, indirectly, about many other movements and interactions that we cannot see.

Shared intentions

Humans do things together and in this sense the sizes of archaeological sites can be informative. Even in Oldowan times, some localities were attractive enough that hominins used and discarded large numbers of tools over long periods. At Olduvai, tools sometimes occur in many successive levels. People were coming back habitually on a considerable scale. The most helpful sites are smaller, often a patch just 5 or 10 m (15 or 30 ft) across. Helene Roche investigated a particularly clear example at Lokalalei 2C in West Turkana, dating to about 2.3 million years. Some of these sites contain splintered bone and evidence of toolmaking. Some 60 or 70 stone cores often make up the source material, each perhaps the size of a fist. As an individual could only carry two or three cores at a time (unless possessing bags that we do not know about) it appears that even these small sites were made by task groups of several people, perhaps bringing in their materials on several separate journeys. This kind of evidence appears, systematically, at just the time that larger brains are also appearing.

From the start of the record, we find archaeological sites associated with animal bones. Even on the earliest sites in Ethiopia, bone fragments are found along with the tools, and the pattern recurs right through the Pleistocene. There is a bias though, because the bones of small animals decay more easily, and bones also have a limited survival time on the ground, so the more and longer a site is used, the less likely they are to be preserved. Even though some other primates hunt small animals, the sharp edges of tools conclusively indicate a far greater interest in meat among the hominins, and the widespread association of stones and bones is very striking. The cutmarks made by stone on bone often survive. Especially remarkable is the number of times that artifacts have been found scattered around a single carcass, usually a large mammal such as an elephant or hippo. Archaeologists have long appreciated the danger that bones resulting from a natural death can come into association with tools through coincidence, in a natural palimpsest. But in numbers of cases sprinkled across Africa one picture recurs: people used stone tools to butcher a single carcass, often of a large animal; one particularly clear example is the elephant skeleton at Barogali in Northeast Africa, dating to 1.6–1.3 million years ago.

An outstanding puzzle is the extent to which hominins hunted, rather than merely scavenged meat. The issues have been debated for more than 30 years, as we shall see in Chapter 6. Within the last 500,000 years there seems little doubt that hunting took place: wooden spears and the sheer concentrations of bones testify to it. But in earlier times, we have

Figure 4.7: *Early* Homo *at some stage learned to hunt large animals. Outwitting the natural defences that make animals such as these Blesbok perpetually wary is essentially a social enterprise.*

difficulty in seeing how small and small-brained hominins could have achieved what modern hunter-gatherers sometimes find difficult.

Perhaps we can bypass the problem, for chimpanzees certainly hunt small animals and baby animals and there are indications from Kanam in Kenya that early hominins too sometimes favoured the smaller animals. Cooperative hunting is attested in chimpanzees, and similar collaboration and 'shared intention' can be inferred in hominins from the clustering of tools, and the extent of carrying materials across the landscape. Indeed we can see that essential cooperation more clearly than we can see the hunting itself.

Why none of this is easy: the problems of visibility

Apart from the ever-present stones, the sobering fact is that most other technologies do not survive in the archaeological record of the Palaeolithic. Fire excels at this vanishing act, but other materials are nearly as successful. As most ape tools are made of soft materials – stems or leaves – it seems a good bet that hominins would have included similar things in their repertoire. Chimpanzees, as Adrian Kortlandt found, will belabour a stuffed leopard with a stick (and Bill McGrew regards chimpanzee use of weapons as well demonstrated, but too little known). Robin Crompton has shown that modern humans are geared to be able to carry a small load, around half a kilo, in each hand. It is hard then to imagine that wooden tools were not being used, but the chances of them surviving are infinitesimal. However, three early cases do defy the odds: wooden tools do survive, around 700,000 years ago in Israel, and at about 400,000 years ago in southern Africa and in Germany. These examples were made by large-brained hominins, but it seems likely that the use of wooden tools goes back a good deal further. For example, Lucy's long thumb would have been useful for grasping a wooden stave.

But archaeology does not have magical abilities to make a rule book for inferring the existence of objects and activities beyond what we recover. What it can do is set out hypotheses that it would be realistic to test, ideally first setting out their limits. Bone tools offer a good example – a handaxe of bone exists from Olduvai, and numerous others from Italy. So we know that bone tools were made more than half a million years ago, and we can infer that they could have been made as far back as the first bone debris found with stone tools – 2.6 million years. So if we want to ask, 'when was the first bone tool made?', there is a huge range of doubt left to us – from about 2.6 to 0.8 million years. In human evolution such a range creates controversy and we must always be wary of over-interpreting evidence.

There is certainty, however, that the early hominins passed through profound biological changes under heavy selection pressures. Although

Figure 4.8: *Preserved wooden spears from Schöningen in Germany, around 300,000 years old, are a rare instance of preservation of wood, but such technology must have been ubiquitous. The finds are a reminder of how much is missing from the archaeological record.*

apes have changed through the last 7 million years or so, our ancestors changed much more – and that is why we are no longer apes. Evolution does not of necessity have a single direction. The very large teeth of 2 to 3 million years ago show the pressures of dietary stress. Subsequently, brain came to pay off more than teeth. Beyond these obvious signs, the social structures and behaviour of hominins were certainly affected. By 2 million years ago we can be sure of the 'wind of change', with geographic distribution, tools and brains all showing firm signs of hominin success. It is perhaps remarkable that the first hominin sociality, of Ardi and the australopithecines, had not required a larger brain. Following the insights of the social brain, we may conclude that up to this point the biological and social changes had happened without fundamental changes to population structures and community sizes. Amid continued pressures, those larger brains were to become an expensive necessity.

Summary

In this chapter we have explored some of the intangibles of human evolution, community size and social structure, and also examined the tangibles of fossil skulls and stone tools. The social brain adds a perspective on the former that cannot be garnered directly from the hard evidence of bone and stone. But you will also see that we have not yet opened the full box of tricks. At this stage our investigation of the deep-ancestry of our social lives has been largely comparative, drawing on studies among living primates and extrapolating them to fossils. We have not mentioned theory of mind, degrees of intentionality or the amplification of social life using those core resources of materials and senses (see Figure 3.2). There is a good reason for that. Although brain sizes in the earliest hominins exceeded the ape/Ardi threshold of 400 cc during the timeframes of this chapter, they were still small-brained hominins. Their social lives were more complicated than those below the 400 cc threshold. But by comparison with what came later, the journey had only just begun. What we have seen are the consequences of bipedalism, the imperatives of defence against predators, the impact on diet of gut reduction and the experiments with social learning as applied to making and using things. With this platform in place we can now turn to the evidence for three pivotal changes that lengthened the social day, transformed grooming and ushered in changes in technology.

5

Building the human niche: three crucial skills

Know yourself and know your place

What really makes us human? One of the big problems in studying human evolution is that *we* define *ourselves*. Some 300 years ago Linnaeus, the great Swedish classifier of animals and plants, labelled us as *Homo sapiens*, the wise one. He also commented '*Homo. Nosce te ipsum*' ('Man. Know yourself'), taking up an ancient Greek maxim with a new twist: we had to know ourselves as a species. Scientifically we are both judge and jury of who is and who is not a member of the human club, a distinction that is not found in nature, but comes with all the baggage of cultural prejudice and historical interpretation. Archaeologists somehow have to decide who makes the cut among our hominin ancestors. They tend to do this by drawing up checklists of items such as ornaments and art as well as activities that include hunting and religion. The problem is that in deep-history we have very short lists of features to examine. And since we have no independent court to adjudicate our claims, it is hardly surprising that there is little agreement on what exactly made us human, let alone when such aspects first appeared.

In this chapter we look at three key elements that are on everyone's list – the distinctive stone tools known as handaxes, fire and language (starting, funnily enough, from laughter). We have selected them for good

reasons. From a timeline perspective, they carry us through the journey from small- to large-brained ancestors, and they all appear, or seem to appear, when, in Africa at least, *Homo* was not yet alone, because the robust australopithecines still survived (see Figure 5.1). These three elements pose very different challenges – handaxes we can certainly find; fire we can find with a great deal of luck; but language itself is irrevocably gone, to be inferred only from anatomy or symbols. But all are the focus for skills that persist and evolve. Fire allowed hominins to transform materials, whether it was roasting meat or hardening the tips of spears; it also changed the mood of those places and times of day when people gathered, providing a boost to the business of interaction. Language advanced the skills of listening and communication and further changed the basis of social life. Speech also played a pivotal role in the development of the mentalizing skill of reading another's thoughts to predict their actions. And, surviving valiantly when almost all else disappears, handaxes reveal more to the archaeologist than simply the skill of

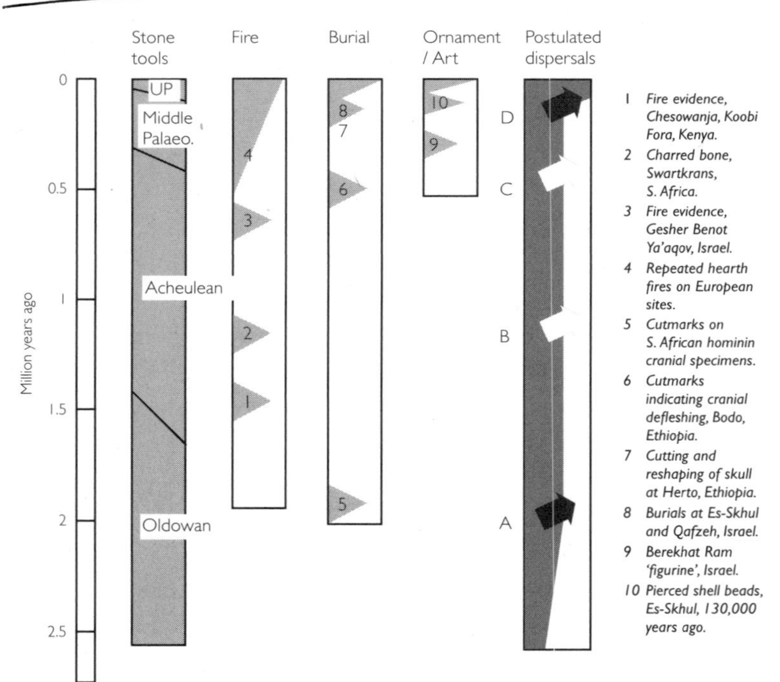

Figure 5.1: *Major divisions and events in the Pleistocene timescale. The postulated dispersals are: A – first diaspora across Eurasia; B – first movements into Europe; C – possible movements of* Homo heidelbergensis; *D – dispersal of anatomically modern humans.*

making stone bifaces – they indicate levels of attention, concentration and precision not found in any other toolmaking animal.

All three elements, and the skills we can infer from their appearance, shed light on the evolving hominin niche in a long but critical period, from 2 million to 0.5 million years ago. From our social brain perspective, these skills evolved as responses to the selection for more intricate and complex social worlds: the difference between the scale of a primate's social world and the challenges of doubling the numbers to live, as we do, in a world of Dunbar's number of 150. As hominins built their niche-for-living, their social community, in ever more complex ways, we believe the three elements are closely linked – a link based on the co-evolution of brains, bodies, materials and surroundings. The result by 1 million years ago was a highly distinctive niche that was home in the following half-million years to the appearance of large-brained hominins with complex skills (see Chapter 6). And because it was our social lives that drove the increase in our brain size and the use of the materials around us, we see these advances as examples of a social technology. Yes, handaxes were efficient at cutting up carcasses and fire kept people warm; but beyond these obvious functions lay a deeper evolutionary drive, to know others and by doing so to know better how to play the social game that led, in the long term, to evolutionary success.

In this chapter we will be following the early trail from hominin to human and completing step six on the timeline in Table 1.2. This is the story of large-brained hominins, even though the smaller-brained australopithecines were still present in Africa until 1.3 million years ago. The key species is *Homo erectus*, remarkably widespread across the Old World (see box overleaf). By the end of step six we find *Homo heidelbergensis* across Africa and Europe and the stage is set for the appearance of the first modern humans.

Handaxes: learning a craft

Sometimes a symbol is so powerful that we call it iconic – like an apple on a computer, or a name on a watch. The first of these great signs (and we do not even know that it was ever intended as a symbol) is the Acheulean handaxe. Today it is the icon of the Stone Age and still widely used, for instance as the inspiration for the logo of the UK's prestigious Royal Academy of Engineering. The handaxe is extraordinarily recognizable

Bridging the brains: *Homo erectus*

In the complex record of human evolution one character has always stood out as steadfast and reliable: *Homo erectus*, who stood tall and rugged and went on and on, bestriding the world's geography. In a general sense, these were creatures who were clearly human, seemed to stay much the same for a very long period, and whose brains were substantially smaller than ours, but significantly bigger than those of the earliest hominins. This picture still makes a good starting point for looking at the last 2 million years, but the more we learn, the more the details dissolve kaleidoscopically into new vignettes, showing us local varieties of hominin whose brain sizes and body frames actually vary markedly.

The story now starts with specimens sufficiently different from the 'traditional' *Homo erectus* first found in the Far East that they even sometimes get their own separate names, *Homo ergaster* in Africa and *Homo georgicus* at Dmanisi in Georgia. Both have can have quite small brains – indeed those from Dmanisi are less than 700 cc, smaller brains than in earlier *Homo* in Africa. African specimens have near-modern body proportions by about 1.5 million years ago but the Dmanisi hominins have hind-limbs that are more transitional in character. If we do class these types as *Homo erectus*, then across a million years the species is no longer the same, same and same again. Rather it is embracing a mosaic of local changes, so that bodies might be stockier here, slenderer there, smaller brains in one place and larger ones somewhere else.

If there is a surprise, it is that far-scattered finds can be so similar. It is easy to suggest that *Homo erectus* spread all the way across Africa and Eurasia, but in fact there is the most enormous gap in our knowledge, all the way from Dmanisi to China, some 8000 km (5000 miles) without a skull. *Homo erectus* was first found in the Far East, and it is from Java and China that our traditional impression comes: rugged humans with brains of around 1000 cc, stockily built, and with very thick cranial bones. Finds scattered around Africa, including a skullcap from Olduvai Gorge, suggest that *Homo erectus* there eventually became rather more like the Far Eastern populations by around 1 million years ago. But it is in Africa, probably, that we are to look for the roots of a new species – the ancestors of *Homo heidelbergensis*, who were appearing by 600,000 years ago at the latest.

Figure 5.2: *Handaxes were hand tools. Their basic functionality was probably the main reason for their long endurance.*

and extremely useful, if somewhat puzzling, because of all the trails that it leaves back towards the intentions of its makers.

Handaxes emerged in Africa by at least 1.75 million years ago at the sites of Lokalalei on the west side of Lake Turkana and Konso-Gardula in Ethiopia; the last ones date to about 150,000 years ago. Sometimes they are found in huge accumulations, as at Olorgesailie or Kilombe in Kenya, but elsewhere there are no more than rare individual specimens. They were successful enough to be used for a million-and-a-half years and their finds extend across most of the Old World, but become rare or absent in parts of the Far East. The classic handaxe is a stone object, perhaps 15 cm (6 in.) long, and about the width of a palm. It often resembles an almond or boat in its outline, pointed at the tip, and rounded at the butt. In the stone knapping process the makers flaked most handaxes from both sides to make a sharp, well-supported edge all around the perimeter. They have an average weight of about 0.5 kg (1 lb), much the same as a tin of beans or small bag of sugar. This may be the typical textbook specimen, but handaxes vary in surprising and intriguing patterns. Even a small assemblage will include shorter and longer, broader and narrower, thicker and thinner specimens. These variations are so large that they have to be deliberate – a big handaxe may be ten times heavier than a small one, and we hardly ever find groups of handaxes together that do not embrace difference.

What can we learn about handaxes by approaching them with the model of the social brain? The problem is that we still don't know exactly what the tools were used for – some would be good for cutting, others for pounding. There could be both general-purpose uses, the Swiss-army-knife principle championed by Steven Mithen, or more specialist uses. One thing we do know: handaxes never formed the whole toolkit.

The big handaxe accumulations show us organization in the landscape. They suggest that hominins were mapped to their landscapes through rule systems that had to be obeyed, that gave guidance in much

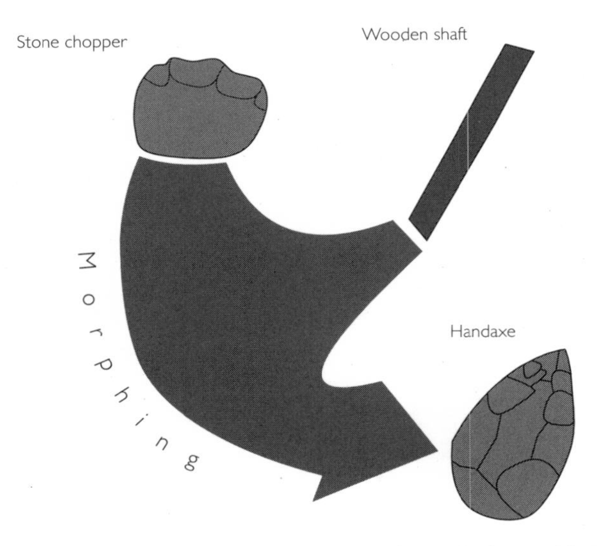

Figure 5.3: *More elaborate tools often represent the coming together of ideas that were already separately in circulation. This can be suggested for the Acheulean bifaces: the implication is a reservoir of socially mediated knowledge that allowed a new flexibility.*

the same way that a powerful religion can prescribe every facet of life for some people today. This was the case at the remarkably well-preserved site of Boxgrove, 500,000 years old, located on the Sussex coastal plain in southern England. Investigators Matt Pope and Mark Roberts were able to show repeated patterns of local landscape use by a large-brained hominin, *Homo heidelbergensis*: finding suitable flints in a collapsing sea cliff, carrying them onto the plain, making distinctive ovate-shaped handaxes, using them to butcher animals and then – and this is the fascinating twist – throwing them away in specially designated places.

Alain Tuffreau, an excavator in northern France, makes a similar point about the Somme Valley, where he has continued research that began in the 19th century. What he found was that the same activities, the same kinds of toolkit, would occur in the same location time and again, even when there were major breaks in the record: 50,000 or 100,000 years later, the same patterns would be resumed. We find similar indications in East Africa that the huge occurrences of handaxes occur only in particular situations. Where they arose, the same repeated patterns would be set up, and go on and on. At Olorgesailie in Kenya, Rick Potts found through extensive exploration that the area where there were many handaxes represented one unique concentration in the whole of the

Figure 5.4: *The famous Catwalk site at Olorgesailie in southern Kenya preserves one of the best-known concentrations of handaxes in the world. Many thousand handaxes and cleavers are exposed on the surface and these were accumulated over a long period of time.*

surrounding landscape. John Gowlett has found the same at Kilombe, further north in Kenya, where the handaxes extend for 200 m (650 ft) along one archaeological surface, but are rare in adjacent areas.

The makers of these tools, *Homo erectus* and *Homo heidelbergensis*, had brains in the range 800–1200 cc, thus showing an encephalization up to twice that of early *Homo*; the hominins at the Somme and Boxgrove had almost certainly crossed the 900 cc threshold. But handaxes were not the sole preserve of large-brained hominins. They are not a reliable marker that brain growth has taken place, neither are they any certain indicator of language. Their symmetry and fine working is often extraordinary; but their time-range travels from small-brained hominins, whose group size was still manageable without vocal grooming, to large-brained hominins, where it was not.

The social lives of stone tools

The implication we draw from these artifacts is that life was lived on a larger scale, a larger scale of mind, involving larger groups and communities who could transmit and maintain more information. The organization of the Acheulean tallies with this. A handaxe appears to us as a single object, but it is actually a node in an extended network.

People might transport material for handaxes for some 10 km (6 miles), and individual specimens appear more than 100 km (60 miles) from their source rocks.

The handaxes are very useful to researchers because they are the first set of tools that embody complex rules. That is why they are so recognizable to us, so widely across the world, because the rules are so similar. We can see that the handaxe is one of the first real inventions, but that does not mean that it was designed by an individual. Rather, it seems a set of ideas came together through a long process of small, perhaps unintentional, experiments and adaptations. Some of the earliest examples lack some of the classic features, but when these dropped into place, they were maintained and maintained and maintained over 1.5 million years.

We would argue that this was possible because there was a firmer, larger, social framework than was available to the australopithecines or apes. This bigger frame was probably crucial as it meant that enough knowledge could be held by enough individuals that it was always reliably available. This would not be true in an ape society, where the community is everything, and cultural contacts have to filter across its boundaries through thin connections, without the help of language.

Archaeologist Stephen Lycett has studied the nature of cultural transmission in the Acheulean, building on a set of ideas previously used by the well-known geneticist Luigi Luca Cavalli-Sforza. The question is, where do individuals get the ideas that they use in culture, and why do these stay the same or change? What we learn is preserved in our memories, and unlike genetics can be changed by re-learning, but it can still become very fixed. With handaxes the puzzle is that they are so much the same *and* different. The rules seem to be that if you are a handaxe-maker, (1) you always follow the basic rules, and (2) you are allowed to vary them in 'sliding adjustments' almost as far as you like. But reality will come crashing in if you push them too far – make it too long, or too thin, and it will break. We do not know if the limits were learnt personally or by tradition, but Cavalli-Sforza and mathematician–biologist Marcus Feldman do lay out the main possibilities: the learning might be vertical, from parent to offspring; horizontal, learnt from a peer group; one-to-many, where a teacher has respect; and many-to-one, where a peer group of elders lays down the practice. These have different results in terms of fidelity of copying, and rates of change. The anthropologist

Dietrich Stout records the importance of 'masters' in stoneworking among the Langda of New Guinea, with novices under strict control. Stephen Lycett believes that vertical transmission has been assumed for the Acheulean, but that many-to-one would in fact best describe what we find (see Figure 5.5). In addition, factors may be at work which take the Acheulean outside modern experience. Glynn Isaac argued that the Acheulean handaxes showed the presence of local craft traditions, with local characteristics, but he thought that the overall similarities in them over huge areas showed an ease of communication flow greater than in modern societies (where artifacts such as pots or ceramics vary enormously in design and decoration). He suspected that earlier, simpler languages were less of an insulating barrier between neighbouring groups than they became in later times.

Here the social brain helps us to think about the numbers. Compared with apes or australopithecines, individuals must have been open to learning from more individuals, with greater flows of information across distance. In a hunter-gatherer band of say 30 individuals, a person learning as a child might well be influenced predominantly by about six others (incidentally often about the number of an ape or human small task group, as also predicted by the social brain). But in a

Figure 5.5: *Social transmission can be seen as vertical or horizontal (e.g. from parent to offspring, or from peer to peer). In the Acheulean, the individual learnt from probably only a few others, but the fidelity of transmission was remarkable.*

lifetime each individual is also likely to migrate between bands, so that knowledge will be exchanged over larger areas.

The tools themselves attest directly to longer-distance movements. On almost every large Acheulean site some of the tools have been carried in from great distances, 50–100 km (30–60 miles). How this happened is not exactly known, but the two main possibilities are that people wandered, or that there were exchanges. Each of these must make us think of the travels, the groups and the group sizes. The distances seem too great to be forced by subsistence necessity. At Gadeb in Ethiopia, J. Desmond Clark found that several handaxes of obsidian had been brought more than 100 km (60 miles) from the Rift Valley. Whether it was straightforward journeying, or the beginnings of exchange, social contacts would be crucial either way, as you would have to engage with people from further away. The bigger brains of *Homo erectus* are a good sign that this kind of social mapping was going on.

Children would probably learn their stone knapping in the first few years (there is anatomical evidence that in *Homo erectus* the progression to adulthood was still fairly fast). We do not know if there is such a thing as a toy handaxe, but modern hunter-gatherer children do have toy bows, which can be used to good effect. In the nature of the band they would get their key ideas and experience from a few others. In the learning there could be strong messages enforced of 'do it like this'. Here we are making an inference, but it is based on the very repetitive evidence that we see. In this respect other primates do not help us much: in chimpanzees there is little teaching and that is generally from mother to infant. Younger chimps often spend time in sibling groups, and in task groups like boundary patrols, learning and teaching must be between adults. The socializing between hominin males may be at the heart of the new possibilities in becoming human, and the number of stone cores transported almost certainly indicates multi-male cooperation.

The power of concentration

So what ideas were transmitted, and what else can we learn from them? It takes probably at least ten ideas or concepts to come together to make a handaxe. We know from other stone tools, and later on from wood or bone, that the concepts can be handled in similar or different combinations in other tools. For example a spear, like a handaxe, presents its maker with challenging judgments about length, breadth, thickness and balance. Of course a spear that is round in section is embodying slightly different concepts than a broad tool such as a handaxe, and that is exactly the point.

Handling such groups of concepts is hard work, even for modern humans. The handaxe-makers evolved sequences of operations, such that they would not normally need to be considering more than three or four factors at a time. Their tasks required concentration, in a way that is echoed by the practice of social skills –in which greater attention is paid to others. We would argue that this concentration was first directed towards the business of building stronger bonds in increasingly large groups. But longer attention spans also made it possible to refine technology by devoting more time to its manufacture and its form. Chimpanzees are notoriously poor at concentration (except in prolonged bouts of gaining food), but humans can spend hours gazing into each other's eyes, solving a problem or listening to a sermon. What we conclude is that greater skill in toolmaking, as seen in the shaping and refinement of implements such as handaxes, was a skill inseparable from the social lives of hominins. The prime directive, as we would expect with a social brain, were the skills that supported our increasingly complex social lives and the pressure of larger numbers. Some of the same skills, however, were equally available for making tools of ever-increasing elaboration and refinement.

Did making handaxes require language for passing on what needed to be learned? We would say not necessarily, but we come to that below. However, it is likely that, when language grew in importance as a grooming mechanism, attention levels would be further increased as hearing replaced touch in bonding. The hominin niche was being transformed by the social brain.

Carrying the cognitive load

Making handaxes is not an easy process. They can be made in varied ways, but normally in three main stages – first selecting a raw material, second obtaining a suitable blank, and third doing the final trimming. These steps may be separated geographically, as the toolmakers parcelled out their ideas through time and space, a very human characteristic.

Usually in Africa the first two stages would happen in one place, because the source would be a boulder too heavy to move far. The standard 'biface' blank would be made by striking off a large stone flake perhaps 20 cm (8 in.) long (biface, the French term, refers conveniently to both handaxes and cleavers). Here we should admire *Homo erectus* for the skill needed in administering such blows.

Of course, the more 'correct' the struck blank was, the less final adjustment or trimming was necessary. Louis Leakey long ago noted examples at Kariandusi where only one side of the biface needed any trimming. Our analyses show that the same basic factors were controlled again and again. The difficulty for the maker is to get them all right at the same time. The problem will be familiar to any modern practitioner of DIY who carefully adjusts one thing, only to find that it puts something else a fraction out.

Acheulean cleavers presented a greater challenge than the handaxes, because although they have the same basic shape the axe edge has to be formed at the same time as the blank – in 'true' cleavers it is not made by later trimming. Acheulean makers seem to have been agreed about the need to achieve this edge, although they achieved similar results by different complicated processes in different places. In the procedure used in East Africa the makers of a cleaver were controlling four concepts at once as they prepared to strike the blank flake – length, breadth, breadth of the cutting edge, and also thickness. Careful preparation of the core would allow this, and then the result would be released through precisely directed application of a carefully weighted blow. All this requires a kind of mental juggling that we more normally associate with modern humans. Generally we ourselves find it difficult to handle more than three variables at the same time. The trick in the Acheulean was to reduce the cognitive demands by careful sequencing of operations. This too occurs in many present day technical processes. In the Acheulean of southern Africa a somewhat idiosyncratic technique of making bifaces is called the 'Victoria West'. It shows a great concern with preforming the shape of the final tool, because a core was carefully flaked not much bigger than the tool itself, and then with the final blow the near-finished item was detached.

If these rather technical procedures can seem a bit much for us, it is good to remember that *Homo erectus* was up to them. The Victoria West technique demonstrates a particular kind of insight on their part – that

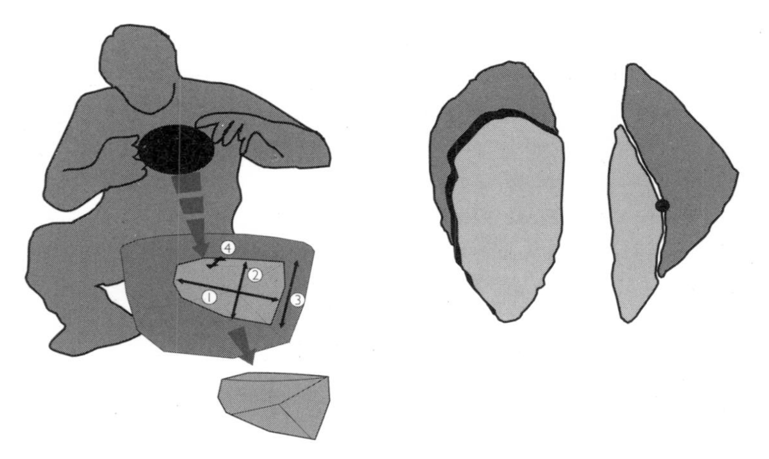

Figure 5.6: *Left: The process of producing an Acheulean biface blank from a large core. Through careful preparation the maker could control several dimensions of the blank (labelled here 1–4) in a single blow. The knapper would naturally work from the near side of the core, and here has been transposed for clarity. Right: In the Victoria West technique used in South Africa the core was often little larger than the biface that it produced.*

the whole shaping process makes no sense, unless one knows what is going to happen at the end.

Now, indisputably, the account just given is 'technical'. But we would argue that the manipulation of multiple concepts in the head is not purely technical. In the first place, social judgments are made at each stage of the process. Shall we go to the source today or tomorrow? Will that interfere with the other things we have planned? Will we take some food with us (in an era before sandwiches), or shall we forage on the way? How many of us are needed to carry the blanks back? Who will guard the children while we are away? Who might we meet on the way? And then too, the co-manipulation of several concepts seems to echo quite closely the kind of levels of intentionality which we discussed in Chapter 2.

Fire: a social history

Fire is a phenomenon of nature that has played a colossal role in shaping human societies. There is little that we could do without controlling heat, although over the last century we have largely excluded raw fire from western homes. Although we have every incentive to explore fire history, in practice it is difficult to do so.

Fire use presents an even more tantalizing problem. Only humans can control fire, and every living society has been able to do so, although not all of them make a practice of kindling it. As a result fire-making is regarded as a sacred activity by some groups – for example the San of southern Africa – so possibly not every ethnographer who asked 'show me how you make fire' gets a direct answer. Undoubtedly, however, fire starts as a natural resource, which humans could come to use if they could find the benefits of natural fires.

There is much debate about how and when this happened. For some archaeologists it is about a 'cognitive leap' that saw the potential of fire, a vision measured by the absence and then presence of hearths in early human settlements. But as fire was first of all a phenomenon of nature, we argue that its use could have begun gradually, and was expanded over time. By the time that it was carried into settlements and made at will again as 'home fire', fire use could already have had a very long history.

The case for cooking

This broader perspective allows us to argue that fire has had a crucial role both in shaping human diet and in building our large social brains. Taking diet first, the apes give us some clue about ancestral food resources. For chimpanzees, fruit is the main source of food energy, and it is largely carbohydrate. They eat less nutritious plants as fall-back food, and broaden their diet with insects, small animals such as lizards, honey, and the occasional, but regular, hunting of mammals such as monkeys and baby antelope. In this way they take in small but significant supplements of protein and fats. *Ardipithecus*, it seems, was more of an omnivore (see Chapter 4), and in the australopithecines we see a definite broadening of diet.

Whereas chimps, gorillas and orang utans all eat or otherwise use hundreds of species of green plants, the variety of resources decreases outside the forests, and away from the tropics. In dry seasons far less is available, and dietary stress can arise very easily. Studies by Eileen O'Brien, Charles Peters and Richard Wrangham demonstrate the importance in these circumstances of other carbohydrates, such as roots and tubers, known collectively as underground storage organs (USOs).

In an important paper of the early 1980s, Peters and O'Brien looked at plant food resources – totalling 461 genera, with many more species – across east and southern Africa, comparing their use by chimpanzees, baboons and modern humans. There were considerable overlaps, but it was plain that all three primate species use great numbers of

Food item	Humans N (%)	Chimpanzees N (%)	Baboons N (%)
Flowers/inflorescences	2 (2)	1 (2)	5 (9)
Fruits	38 (41)	34 (72)	14 (26)
Seeds/pods	9 (10)	3 (6)	11 (20)
Leaves/shoots	22 (24)	7 (15)	15 (28)
Stems/stalks	4 (4)	2 (4)	1 (2)
Underground storage organs	17 (18)	– (–)	8 (15)
Totals	92 (100)	47 (100)	54 (100)

Table 5.1: *Plant foods consumed by human hunter-gatherers, chimpanzees and baboons in central and southern Africa (N = numbers of plant genera used). There are considerable overlaps, but underground storage organs (USOs) are not used by chimpanzee fruit specialists, and humans use more kinds of USOs even than baboons, probably because cooking aids their digestibility.*

plant species, but that chimpanzees generally do not use USOs, while both baboons and modern humans do (see Table 5.1). Unfortunately these foods are often less digestible, both for modern humans and their ancestors.

We are not good at digesting the starches found in tubers, and excesses of meat protein can even poison us. A vital advantage for fire thus comes into play: cooking breaks down the structures of food components – the packages of starch in tubers and the strands of protein in meat – so that they can be more easily dealt with in the gut. It also destroys noxious microorganisms. Of course early hominins would not know these things. What they would find in the first place was that foraging near naturally occurring fires would make available more foods, and that those foods accidentally cooked were made more palatable. Tubers and burrowing animals may have been the most obvious targets, as both would be more visible once a fire had passed. Ethnographic accounts tell us much of this for modern human foragers, and in this case we cannot doubt that the same things were true in the past. Nor do we need to start with a cognitive leap, because many other species, birds in particular, are known as 'fire followers'. They show that fire occurs regularly enough in nature to be exploited in patterns of learned activity. It is not surprising that archaeologists living in green and pleasant lands should see fire as a rarity, but quantitatively large numbers of lightning strikes occur each year, at least from the equator to 50 degrees north or south, and a measurable proportion of the ground strikes causes fires. These occur particularly where the vegetation is dry, and when lightning strikes get ahead of the heavy rain that often accompanies them. The rate and scale of fires is very much determined by local factors. In tropical rainforests, nearly everything is too wet to burn – but in semi-arid zone grasslands, fires occur so regularly that many tree species are fire-resistant. In temperate woodlands and forests, fires are a risk every summer, particularly in times of 'fire weather'.

When would this kind of encounter between hominins and natural fires start? Logically it would come when they began to range more widely in environments that had regular burns. The stone tools show just such a scale of foraging around 2.0–2.6 million years ago, in environments where the presence of fire would have been unmistakable; nowadays it can be seen from more than 20 km (12 miles), by night and by day (see Figure 5.7).

Modern human teeth still reflect some of the changes in diet, in an unbroken line with the past: our molars in particular show the need to grind hard or tough foods. Ours are not predominantly cutting teeth, because food preparation has been delegated to tools such as knives. One find gives a hint of how early such changes may have happened: a cranium of early *Homo* from Dmanisi in Georgia, aged 1.7 million

Figure 5.7: *Bushfires are highly visible on the landscape. This one, seen from Kilombe in Kenya, is prominent at about 15 km (9 miles), and was equally visible at night from its glow. Natural fires are a feature of life at low latitudes and our earliest ancestors would have been familiar with them.*

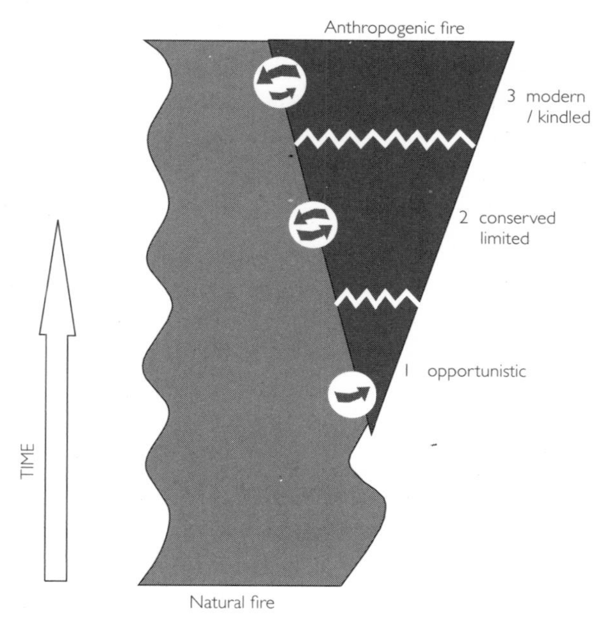

Figure 5.8: *A general outline for the development of fire use, showing its emergence from and interchanges with natural (wild) fire. Three stages are recognized, each showing more control by hominins over this vital technology.*

years, completely lacks teeth except for one in the lower jaw. 'Old toothless' had survived, probably for several years as the healing of the tooth cavities indicates, certainly because his food had been adequately prepared, and very likely because he was a member of a social group in which others provided support.

Does this mean that actual fire use goes back so far? There is no evidence of it at Dmanisi. A problem to reckon with is that early humans now appear to have spread out right across the Old World – from Africa to Georgia, to China and Java – before their brains grew larger, as a skull published in 2013 has emphasized. The distribution implies that the roots of this dispersal (and a new successful *Homo* adaptation) may go back before any claims of fire.

Richard Wrangham and his colleagues nevertheless believe that fire had a strong early impact. They have set out arguments to show that hominins could only have survived on the savannah and through dry seasons by substantially altering their diet, and in particular by learning to cook meat and carbohydrates. They point out that all modern humans need cooked food, and that health deteriorates rapidly without it.

Certainly, the increase in brain size of some 50 per cent by 1.7 million years ago is something that needed to be fuelled. Could it have been achieved simply by improvements in quality of diet that were themselves made possible by a more supportive social world? Some major dietary issue must have been solved, or *Homo* could not have done without the earlier megadonty (the massive teeth characteristic of many australopithecines). Human brain growth occurs early in life, so central factors would be the health of lactating mothers, and the ability to wean early to high-quality foods. The old man of Dmanisi shows again that the problem of food preparation had been solved, but was it by fire or by social collaboration and the careful preparation of food, aided by cutting and pounding technologies? Writing together in 2013 John Gowlett and Richard Wrangham tried hard to bridge the gaps between positions. Both are convinced that fire was in use at an early date, but as with other developments, the hardest thing for us to imagine is the form of adaptations that were different from modern human uses and behaviour to the point of being beyond easy analogy.

Firing up

The archaeological evidence for fire points to a slow burn. The rub is that archaeologists must begin by looking for evidence that is not actually the first indicator of fire control. Hearths in settlements are common in later archaeology and get progressively rarer as we go back in time. Sometimes the gaps between occurrences are so great that they warn us to be careful about over-interpretation. They show that most fires do not get preserved at all, even long after we know fire use was well established. The oldest traces of fire that could represent hearths come from three sites in East Africa, Chesowanja, Koobi Fora and Gadeb. All are regularly cited, and then usually dismissed as inconclusive. At Koobi Fora and Chesowanja at least, burning is proven, and the discrete nature of the burning is clear. At Chesowanja, the burning is also demonstrated to be within the range of modern campfire temperatures. The point that is lacking could be called 'triangulation'. If we could show that small numbers of artifacts were burnt, or the bones nearest to the fire were charred, this would create the supporting interconnections of a triangle. As it is, we can just show that burning happened very locally.

The next set of evidence comes from around 700,000–900,000 years ago, and occurs at points from South Africa to Bogatyri by the Black Sea. Best known is Gesher Benot Ya'aqov in Isarel, where people camped beside a small lake about 700,000 years ago. Burning at the site can be convincingly linked with humans because it occurs at many levels, and tiny pieces of burnt flint mark out what appear to be 'ghost hearths'. Forest research data from North America show that lightning fires usually occur up on ridges: repeated fire events close to water, as at Gesher Benot Ya'aqov, fit much more with a human pattern. The two South African sites are in caves, and both are around 800,000–1 million years old. At Swartkrans fragments of burnt bone have been found in many levels of Member 3, some of them bearing butchery traces, and at Wonderwerk Cave one whole level is full of evidence of burning, which microstratigraphic studies show is not consistent with guano fires (which can sometimes occur in caves). Together these two sites provide very good evidence of human fire control, and they also tally with something else, as already noted in Chapter 4: in the lower levels at Swartkrans, and at another cave, Makapansgat, many remains of australopithecines are found – but later on, when stone tools show the caves were occupied, the number of *Homo* remains is very small. We may infer that *Homo* was far safer from predators – and that fire was a main agent in making caves safe to inhabit.

Figure 5.9: *The fire became a focus for both social and economic activities from at least 400,000 years ago and probably far earlier. Fire played a key role in extending the length of the social day and in roasting food for bigger-brained hominins with a reduced gut.*

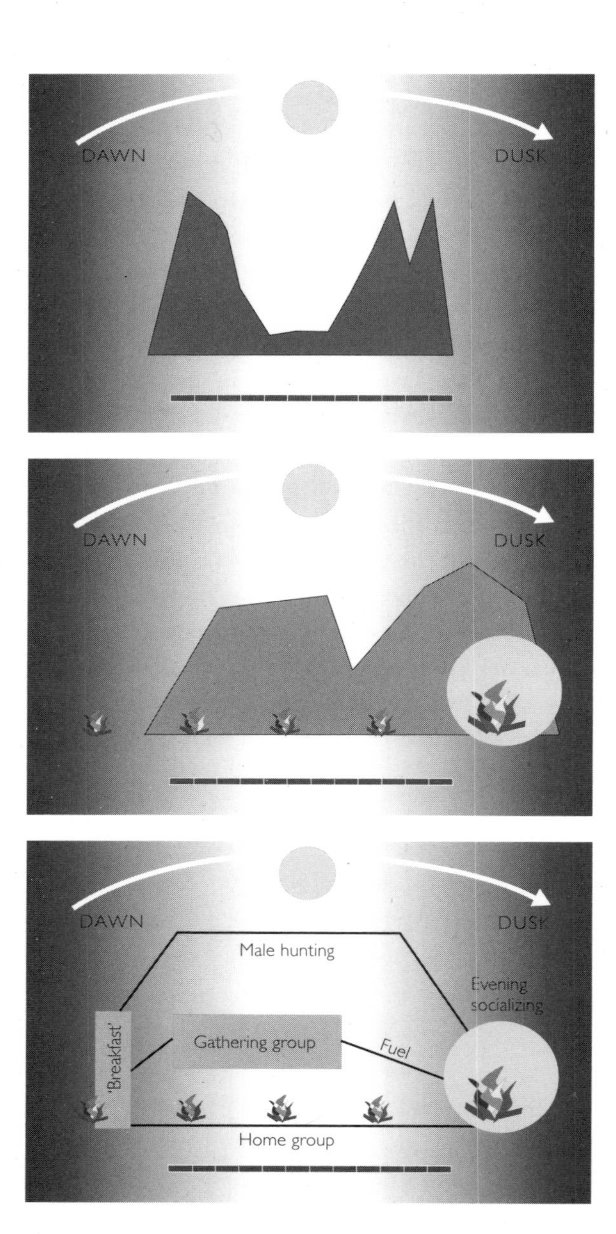

Figure 5.10: *Top: the gorilla day as recorded by biologist George Schaller; the day is limited by daylight and consists largely of foraging periods separated by a midday nap. Centre: the human day has become far longer, peak alertness coming in the early evening, providing extra socializing time that was fuelled by fire. Bottom: Fire has become a factor in the daily dispersal patterns characteristic of hunters and gatherers. (Barscales indicate approximate tropical daylight hours, 0600–1800.)*

If all of this evidence can be doubted, a set of sites in Western Europe, the Middle East and Africa aged around 400,000 years finally gives archaeologists the things that they need to be certain. Schöningen in Germany is renowned for its preserved wooden spears, but also according to its excavator, Hartmut Thieme, has hearths, with one where a wooden stave is abandoned part-burnt. It gives conclusive evidence of human involvement of the most selective kind. We cannot be quite sure that a steak was roasted on the end of the stick, but in every other respect the evidence is definite. At Beeches Pit in eastern England hearths are similarly preserved. Refitting flints (i.e. putting back together the separate flakes) reveals a story of a person sitting by the fire, knapping a handaxe from a large flint nodule. More than 30 blows were struck. Two of the flakes fell forward into the fire and are burnt bright red, while the rest are unaltered.

Fire and the social day

Beeches Pit is particularly illuminating for showing the way that different chains of activity were brought together around a fire – the earliest case yet of something that we see more and more as time went by. Large fires such as those at Beeches Pit take 50–100 kg (110–220 lb) of wood fuel per day, strongly encouraging a division of labour. They indicate that a reorganization of time had already happened (see Figure 5.10).

From the point of view of the social brain, fires tell us something further. An obvious disadvantage of living either in the north or extreme south is that days are far shorter in winter. There is an energy gap: the daylight hours of foraging are fewer, but the need for energy is far greater. Simply put, instead of needing to gain 2000 calories from a 12- to 14-hour day, a human may need 3000 or 4000 from 7 to 8 hours. Humans usually take their meals following preparation, in intensive short periods, often shared – this food-sharing was taken by archaeologist Glynn Isaac to be one of the prime driving forces in human evolution (in fact chimpanzees sometimes share food, but they do not prepare it collectively). Fire seems to have had an enormous part in the change to this pattern, a fundamental restructuring of the daily routine. Further evidence for it is seen in our circadian rhythm: humans sleep a short eight hours, and have their peak alertness in the early evening, just when apes are going to bed. It was fire that made the social day longer, changing the pattern of preparing and taking in food, stimulating and benefiting from divisions of labour. Fire divides labour for efficiency, it enhances the calorific return from food as well as giving warmth and protection from predators. Fire reshaped our day and fuelled the growth of our social brains.

One topical issue is whether fire was universally adopted, or whether it was favoured more in some areas than others. Archaeologists Wil Roebroeks and Paola Villa have argued that fire was not used in the north at early dates. Is it possible that fire was known and managed in the tropics, but not in northern latitudes? The idea is a strange one, because surely fire would be needed more in colder climes, especially in winter – yet the evidence is puzzlingly lacking. At the cave of Arago in the Pyrenees reindeer were being butchered more than 500,000 years ago, but the bones are not burnt, and there are no hearths. There seem two possibilities. One, that fires could be managed only in particular places where fuel and shelter were readily available – in which case we may eventually find the proof; the other, that fire could not be managed well enough for people to rely on it – if fire could not be rekindled, in cold climates lacking in lightning strikes, then it would be very dangerous to come to depend on it (similar risk factors apply to some modern industries). Strong social networks would be required to ensure that you could fetch fire from someone else when you needed it. The strength of networking would be even more important when people were forced to live at low densities on the landscape. In a further Lucy project study Matt Grove was able to corroborate by mathematical modelling that just such factors would indeed apply strongly to people living widely dispersed in rigorous northern latitudes,

Language: the hard evidence

Once we can see people performing activities by the fireside, it is easy to think of communication and conversation. Was language indeed a part of our early picture? Did it shape human evolution far back in time? Linguists, archaeologists and anthropologists have long puzzled over this crucial question. They remain deeply divided. Some argue for beginnings 2 million years ago, while for others language comes only with a symbolic revolution, as witnessed by the explosion of art, about 50,000 years ago.

Our experience of social life is to think at once of language. Surely, then, our social brain perspective can help us in attacking the long-range issue of language origins? We believe it does. The insights into the time-table of speech are one of its major contributions. But first let's look at the hard evidence.

In order to speak, to use language, hominins needed at least the following:

- Brains that could handle the ideas of language – especially syntax (the rules that govern the arrangement of words in a sentence). Syntax is a shorthand way of saying that they understood the complex substitution of sounds to describe and discuss things and concepts, now, in the past and in the future.
- Vocal tracts that could control breathing precisely enough to make the sounds and speeds of speech.
- Something worth talking about to repay the evolutionary effort of re-designing brains and voice boxes.

The first of these requirements is difficult to investigate and divides opinion the most. Internal casts of hominin brain cases – endocasts – give some clues, although most detail is gone, and reading them is somewhat like looking at a face behind a stocking mask. Palaeoanthropologists Ralph Holloway, Phillip Tobias, Dean Falk and others have demonstrated brain reorganization in the australopithecines, compared with the apes. Some of their skulls show signs of greater development in laterality (or sidedness) of the brain, which could relate to tool-using and handedness, but may also point towards the beginnings of language (see box overleaf).

Even so, the brain does not appear to have undergone any fundamental reorganization for language, although some researchers, including Ralph Holloway, have argued on the basis of endocasts, and particularly the position of a fissure named the lunate sulcus, that the proportions of the parts of the brain change between *Australopithecus* and *Homo*.

The mechanics of speech production are only slightly easier to study. Speech requires millisecond-precise timing of airflow from the lungs. The very well-preserved skeleton of a *Homo erectus* boy from Nariokotome in Kenya, dated to 1.5 million years ago, provides solid

Brain laterality, handedness and language

Laterality in the brain refers to the finding that its two halves perform different functions. Handedness is a prime example. Most of us modern humans are fairly definitely either right-handed or left-handed. All around the world the ratio of right-handers to left-handers is about 85:15, and this division is a uniquely human characteristic. It is certainly under genetic control, although its expression can be slightly affected by misguided cultural efforts to suppress left-handedness. Owing to an ancient crossover of functions in the brain it is the left cerebral hemisphere that controls the right hand. The dominant eye tends to match with the hand, and people may be left- or right-footed too.

This handedness may have arisen from the brain's need for concentration in complex tasks. Dividing them between two hands and two hemispheres may have been too high a demand for the millisecond timing and fast responses that are required. This is not to deny that we can divide tasks between the hands at a lower level, for example when digging with a spade, or wielding a stick. Chimpanzees also excel in such tasks, and like us commonly have a preference of which hand to use for which role.

On the whole language too resides in one hemisphere – usually the left hemisphere in right-handed people, but not always the right hemisphere in left-handed people. All this provides a bit of a puzzle – apart from investigating what is going on in modern humans, there is the challenge of working out an evolutionary scenario.

Handedness is the easier part, because it can be traced in the past. The striking of the blows in stone working tends to reflect a right-handed or left-handed pattern of working, and shaped tools too may have the working edge facing a preferred direction. All the major studies carried out so far strongly suggest population right-handedness, whether in modern humans, Neanderthals or earlier *Homo*. Language for many people holds the greater interest, but here things become more difficult. Slight local asymmetries of the brain (petalias) can be measured in some endocasts of fossil hominins. The recognition of development in Broca's area and Wernicke's area (see Figure 2.3 for their location) suggests that language had beginnings as far back as early *Homo*. Even so the position is much less certain than for handedness – and handedness itself need not necessarily indicate the presence of language. It does probably indicate a specialization in concentration-demanding technical processes, and these are indicated by the tools themselves to go back more than 2 million years.

negative evidence. Palaeoanthropologist Anne MacLarnon has shown that he lacked the changes to nerve openings in the vertebrae which would be expected if he could speak. Nariokotome could make sounds but he could not control them as human speech. He was not able to produce long phrases from a single breath and punctuate these with very rapid breaths. This is how we break up the pattern of speech and make it meaningful.

These findings led Nariokotome's finder, Alan Walker, to determine that the boy did not have language. However, at 1.8 million years ago the somewhat older remains of *Homo georgicus* from Dmanisi in Georgia had breathing control that, allegedly, was exactly similar to that of modern humans. But to get conclusive anatomical evidence we have to move forward enormously in time to another remarkable site. Some of the crania from Sima de los Huesos, part of the Atapuerca cave complex in northern Spain, are so well preserved that their bony ear canals survive. These are the hard shells of our outer ear that are tuned to enhance the frequencies of speech, something not found in the chimpanzee. Unfortunately the date of these specimens is not established to everyone's satisfaction. However, it is widely agreed that they are on the evolutionary trajectory from *Homo heidelbergensis* to *Homo neanderthalensis* – at the oldest they could be 500,000 years old, but at the youngest 250,000. Even if the youngest date is right, they are undoubtedly Neanderthal ancestors. The implication is that both Neanderthals and moderns had this speech characteristic, and so either they evolved it separately, or it is part of a background going back to a common ancestor more than half a million years ago. The recent isolation and publication of the Neanderthal genome, obtained by geneticists from DNA in fossil bones, shows that our ancestors diverged from them only during the last 400,000 years. In many ways we have been on the same evolutionary path, but there has also been time for important differences to emerge.

The lack of obvious new features in the anatomical record encourages some to think that language could be largely a learned phenomenon, or even the sudden product of a single mutation, perhaps less than 50,000 years ago. That interpretation seems unlikely. Genetic studies show that some modern human populations had already diverged from one another more than 50,000 years ago, and yet we are essentially the same, using language in the same way. One gene may offer a clue. This is FOXP2, shared with many other species, and generally very 'conservative' – that is, it has changed little over many millions of years, except in humans. In our lineage there have been several key mutations, and it has been found that the gene, which is not a language gene but vital for language, will not allow language to function if it is disabled by mutation. Recovery of the Neanderthal genome has demonstrated that they too had the modern version, in which case our modern version of FOXP2 is also likely to go back to a common ancestor of Neanderthals and moderns at least some 400,000 years ago, and to have been shaped by earlier processes of natural selection.

The social brain hypothesis argues that the great increase in brain size in the Pleistocene was linked with changes in group size. And group size very much determines our need for language. As brain size changed gradually, it is very likely that language evolved gradually. That explains a lot of our difficulty in thinking about it – because we tend to regard language as present or absent, just as we have seen that people tend to think of fire as present or absent.

In the case of language, much of the issue is what would we and they talk about. A common mistake has been to assume that language-is-language-is-language, but for Neanderthals there is every likelihood that they mapped out their world-view with speech in a way that would seem deeply strange and different to us.

Something to talk about

To sum up, the hard evidence for language is patchy and open to several interpretations. However, at some point between the hominins of Nariokotome and the Sima de los Huesos, voice boxes and brains had evolved to the point that some form of language was possible. When in this million years (1.5–0.5 million years ago) is a question that social brain studies can help to address.

Ultimately, the social brain evidence arises out of anatomy. An increase in brain size, particularly the neocortex, is strongly evident through the Palaeolithic – shown in the top graph in Figure 3.4 – and allows us to predict group size. As we have seen in earlier chapters, the 'follow on' from this graph is to interpret the grooming times that are needed in larger groups – it indicates that a hominin with a brain of 900 cc rather than 400 cc had to spend double the time in fingertip grooming the larger number of individuals in their social network. These figures strongly suggest, as Robin Dunbar wrote in *Grooming, Gossip and the Evolution of Language* (1996), that an alternative mode of communication to grooming would be needed at least half a million years ago and probably earlier. In an earlier paper with Leslie Aiello, he argued that there would have been a fundamental shift in the means of communication as the pressure of numbers of interaction partners squeezed the time available. One solution might have been to extend, by managing fire, the length of the social day, putting to another use some of that socially dead time when night fell. We believe this took place, but it was no more than part of the solution. At some point fingertip grooming had to be abandoned as the principal means of interaction between the majority of social partners. In its place we find a form of vocal grooming, but not necessarily language as we know it at first. Indeed, humans are distinctive even now in having remnants of several communication systems – gestures, exclamations (as in 'ouch'), laughter and spoken language. Indeed laughter may have been very important for the evolution of language (see box overleaf).

The big advantage of vocal grooming is that it speeds up interaction. It moves away from touch and the opiate rewards that such grooming releases. Aided by audible communication, an individual could vocally groom an audience while reserving fingertip grooming for only the most intimate of social networks. Basic vocal grooming (however constituted) is also amenable to nuanced amplification through singing and chanting, supported by group activities involving ceremonies, dancing, music-making, laughing and crying.

So what did they talk about? The answer is simple – one another. Humans are built for gossip and there is no reason to believe that this was any different half a million years ago or earlier. Language is our ancestral way of learning about others and influencing them to sign up to our social projects. These can be as mundane as deciding what to eat and with whom. Of course, once language was available it would also have helped in the food quest and potentially transformed social learning about how to do things, and why. It also became a tool for ideology that eventually got thousands to follow particular political colours and articles of faith (see Chapter 7).

The sound of laughter

We share laughter with the great apes, and in particular the chimpanzees, although the form of laughter has been exaggerated in humans. Laughter is essentially a play vocalization in apes, and derives from the typical primate play invitation call. In apes, it consists of a series of alternating exhalation/inhalation bouts based on the natural breathing cycle. In humans, it consists of a series of exhalations, with no inhalations until right at the end when the lungs have been emptied. This capacity to maintain a long series of exhalations is unique to humans, thanks to bipedal locomotion. In quadrupedal animals like monkeys and apes, the shoulder locks the chest wall whenever the weight is on one arm during movement, and this means they can only take one breath per walking cycle. In humans, the arms are freed from weight-bearing, and so we are able to disconnect the breathing and walking cycles. This becomes important later for the evolution of speech, because this too depends on being able to sustain long, uninterrupted exhalations. Otherwise, we would end up with one-word sentences!

Laughter is intensely social and highly contagious. If several other people laugh, it is very difficult not to laugh as well, even when we didn't hear the joke. We think laughter perhaps began as a form of wordless chorusing, long before language evolved. If human laughter evolved out of ape laughter at the very origin of the genus *Homo*, then it might effectively have served to increase the size of community that could be maintained by the endorphin bonding mechanism. In effect, it allowed a form of 'grooming-at-a-distance' that enabled early *Homo* to overcome the constraint on social community size imposed by the fact that grooming is a one-on-one activity. In effect, laughter allowed a one-on-many, or at least a one-on-several, grooming relationship.

The key issue here is just how big a typical laughter group actually is. We tend to think of laughter in terms of stand-up comedy shows – one comedian and a very large audience who are all rolling in the aisles. In fact, when Guillaume Dezecache collected data for the project on this in pubs, he found that laughter groups were close to the limiting size of conversation groups, irrespective of how many people happened to be in the social group. Conversation groups have a natural upper limit at four people (as Robin Dunbar showed some years ago), and the number of people laughing together seemed to have a natural upper limit at about three. This was much smaller than we had expected, but it points to the intimacy of laughter as a behaviour. We are perhaps misled by the fashion for comedy clubs into thinking that many people can laugh together. In fact, casual observation suggests that only rarely does the whole audience at a comedy club laugh together: rather, there are pockets of laughter, often triggered by a handful of individuals in the audience, and these create a kind of vocal Mexican wave in their immediate vicinity. But the effect dissipates quite quickly and the burst of laughter doesn't spread that far.

Three people in a laughter group effectively trebles the grooming group size: I can fingertip groom only one person at a time, but I can make two people laugh and, since I invariably laugh myself, three people get an endorphin kick at the same time. The relationship between grooming clique size and social group size isn't straightforward, but trebling the size of the grooming group would effectively allow you to double the size of the community that could be supported by this mechanism. Being able to double community size from 50 to around 100 is exactly the scale of increase that occurs between the australopithecines and the end of the *Homo ergaster/erectus* lineage.

Speaking your mind: orders of intentionality

We raised the issue of the human skills of mentalizing in earlier chapters, a skill that depends on a concept known as theory of mind. Mentalizing skills comprise what philosophers have termed the 'orders of intentionality', where each level or order of intentionality represents an additional mind being added to the sequence (see Table 5.1). In this sequence, theory of mind (or second order intentionality) represents a crucial rubicon – the ability to realize that another individual has a mind like your own and can believe the things that you believe. The issue we must deal with here is whether it is possible to have a theory of mind without language? In the Lucy project one of our doctoral students, James Cole, researched the issues, looking for the archaeological evidence.

Crucial in the origins of language, it seems, are our abilities to consider and reflect on ourselves (see Table 5.2). Many animals are self-aware. Chimpanzees and elephants can recognize themselves in mirrors; cats and dogs do not. Second order intentionality, being aware of another's mind, is only reached by humans and a few well-tutored, captive chimps. (Dogs may be an exception – when your dog senses that you are miserable, and comes and sits on your foot, is it anthropomorphizing to think that he understands? Daniel Dennett, in his well-known book *Animal Minds*, suggests that dogs are now socialized to an exceptional degree.) It seems likely, judging from their brain size relative to a chimpanzee's, the complexity of their artifacts, and the patterns of their socially mediated activities on landscapes, that the makers of Acheulean handaxes reached a higher order, probably the third.

James Cole argues that the handaxe-makers were unlikely to have had speech based on syntax and its complex use of symbols. But one

Order of intentionality	Achieved by	Examples
Sixth	Only a few modern humans	Complex symbolism
Fifth	Modern humans with language as we know it	Myth and storytelling of increasing complexity and involving real and imaginary worlds and their cast of characters
Fourth	*H. heidelbergensis* and Neanderthals	Shared religious beliefs involving several people and ancestral beings
Third	All large-brained hominins (>900 cc)	You have a belief about her belief which is not my belief
Second (theory of mind)	5-year-old children, all small-brained hominins (400–900 cc) and probably great apes	I have a belief about your belief
First	Monkeys and lesser apes and some mammals such as elephants and dolphins	Self-aware as judged by recognizing yourself in a mirror; a belief about something

Table 5.2: *Orders of intentionality and who achieved them.*

can make a case, as John Gowlett has done, that the recurring concepts found in handaxes may point to an early labelling or indexing with characteristics of language (somehow a great deal of information was being passed from toolmaker to toolmaker). The handaxes involve a complexity of design requiring a mental overview, and as the maker has to integrate the manufacturing steps through time, it can be said that there is a kind of operational syntax. There was certainly some kind of communication centred on paying close attention to others and employing visual cues. Sounds were also likely to have been used to direct attention and express moods, just as they are in monkeys and apes.

We can follow levels of intentionality up towards the fifth order commonly found in modern humans. Third order intentionality marked a step towards language-as-we-know-it. The same mentalizing argument can be applied as with handaxes, but this time to composite tools, as discussed by project member Larry Barham and examined further in Chapter 6.

This type of reasoning applied to technology is comparable to the more recent human practice of creating a universal kinship category such as an 'uncle' or 'aunt', and then applying it to people who are unrelated, to make them 'us' rather than 'them'. Universal kinship is a social skill that anthropologist Alan Barnard sees as deep-seated in hominin ancestry. It allows social categories that do not occur naturally (i.e. genetically) to be created to meet survival needs by the formation of webs of obligation.

Project member Ellie Pearce's brain-size analyses for Neanderthals and earlier humans show rather clearly that if they had language, it could not have been fully modern language – they could only aspire to fourth order intentionality, and that would have meant much reduced language complexity.

The highest order of intentionality (fifth for most people) can be seen as the world of myth, legend and the ancestors. This extension of our social universe in this way allows us to achieve very many forms of intentionality. We suggest that the skills seen in technologies of the fourth order reflect new abilities for coping with absence and separation so that social life continues. Such absences might be in an afterlife as well as overseas or outside the room. When materials are moved and exchanged over great distances, or bodies and objects are made to conform to ways of representation that are acknowledged as the way to do things over a wide geographical area, all speak to the ability keep track of what people are thinking even when we don't see them every day.

So in answer to our question: some form of language, not necessarily speech, was needed if a hominin was to have more than second order intentionality and a formal theory of mind. But as Cole points out, we should not be too rigid in deciding which hominin had what level of mentalizing. Chimpanzees show considerable regional variation in what tools they make – and also (as we have known since the pioneering work of Wolfgang Köhler) much individual variation in their degree of insight. There is no reason to suppose that populations of hominins did not differ in a similar way. It is not possible to say that a widespread species such as *Homo erectus* was capable of third order and only and ever operated at that level. After all, we have seen that Acheulean technologies were made by both small- and large-brained hominins, vocalizing or not as they chipped their way to immortality.

Summary

The fossil evidence points to increasing group size as interaction partners increased in the period from 2 million to 0.5 million years ago. Standing back, we also see that a fundamental long-term shift in the hominins is towards doing things that extend further over space and through time. The scales of society were being extended. We can see this from 2 million years ago. Enhanced communication *must* go back so far and this was manifest in a theory of mind. If language does not leave a major mark on the brain, and the asymmetries of laterality are not pronounced enough to prove major developments, we can at least be sure that the essential machinery, both in terms of cognition and in terms of the anatomy of speech, was being put into place. Altogether we can put together:

- The input–output anatomy.
- The social brain equations.
- The extended landscape demands.
- The concept sets seen in tools.

During this 1.5 million years there is no doubt that early humans were going through very strong selection pressures. There was a driving need for better communication when considering more complex sets of ideas and interactions. All these things militate in favour of a long and slow development of increasingly language-like communication (not a matter of its presence or absence, and at the start certainly not language-as-we-know-it), and the social brain tells us especially about the advantages of getting insights into the minds of others that are expressed in terms of increasing orders of intentionality.

What we have concentrated on in this chapter have been the three key social skills of transforming, communicating and attending (paying attention to what someone else is doing or saying) – skills that apply to others as well as things, to materials as well as the senses that are the resource out of which social bonds are fashioned. Now we will turn to their amplification by large-brained ancestors after half a million years ago. It is time to meet *Homo heidelbergensis*, and its quirky descendants – ourselves and the Neanderthals.

6

Ancestors with large brains

When worlds collided

Imagine this: aliens, driven by long memories and a curiosity characteristic of intergalactic tourists, visit earth every 500,000 years. On their trips 1.5 and 1 million years ago they would have found *Homo erectus* and marvelled at the slowness of change. On their next visit 500,000 years ago they would have met *Homo heidelbergensis* and been largely reminded of the previous two tours, even though hominin heads were now significantly bigger. The last thing they might have expected would be to find us here the next time – urban-living, super-technological *Homo sapiens,* exploring the solar system and messaging on smartphones. They would also be struck by the fact that those wise heads were not much larger than the last time. The 'Nothing much changes' tour now had to be remarketed back home as 'Blink and you'll miss it'.

However we look at it, the acceleration of change has been remarkable, taking us from silica to silicon technologies, fireplace to cyberspace and from a global population of 7 million at the end of the last Ice Age to 7 billion today, and rising. Such change could only happen if the social structures were there to carry it, but what drove the process?

All three species of *Homo* discussed in this chapter – *H. heidelbergensis* and their two closely linked descendants, *H. neanderthalensis* and *H. sapiens* – had large brains. These ranged in size from 1200 to over 1500 cc. All three species are closely related, *H. heidelbergensis* the ancestor of both the Neanderthals of Europe and Western Asia as well as modern humans in Africa, and with a brain size at the lower end of this range, but still large by comparison with the widespread *Homo erectus*. But large brains, and by inference more complex social communities, do not explain the timing of the changes our intermittent alien tourist would have seen. Moreover, size is not everything, as we shall see with the configuration of Neanderthal brains. The social brain graph predicts a group size for the Neanderthals of 140–50 but the configuration of their brains suggests a smaller size of 120, more in line with *H. heidelbergensis*. In this chapter we are examining steps 7 to 9 in the hominin to human story (see Table 1.2).

Turn up the volume

One of the archaeological conundrums that made us scratch our heads during the Lucy project was the lack of innovation among these big-brained hominins. It is not until 50,000 years ago, almost 800,000 years after the appearance of *H. heidelbergensis,* that a whole range of artifacts, loosely grouped under the label of art and ornament, become widespread. There are precursors in Africa, such as simple shell beads from the Cape and Morocco, but they do not sweep the rest of the world until much later (see Figure 6.3). And even they come long after brain size had increased. Surely big, expensive-to-feed brains should have an instant effect on the design of tools and the ability to use artifacts in symbolic ways by exploring higher orders of intentionality? At least that would be the way archaeologists traditionally might expect brain growth to have an impact on culture. So the conundrum remained: big brains, more complex social groups but by and large the same old simple stone technologies. Why didn't we go from the Palaeolithic to the technologies of the present in a fraction of the time that it took?

And then the penny dropped. We are very used to technological and cultural change. It is one of the givens in our world. However, our brains are the same size as a *Homo sapiens*, such as Herto in Ethiopia, who made what archaeologists call a Lower Palaeolithic technology 160,000 years ago. What we have seen in the intervening period is a process of *amplification* applied to technology. This amplification, which we liken to increasing the volume on your headphones, can be measured by the variety and diversity of cultural things and the techniques by which they were made. An excellent example is provided in Brian Fagan's survey of *The Seventy Great Inventions of the Ancient World* (2004), which begins with stone tools and ends more than 2 million years later with contraceptives and aphrodisiacs dated to 1400 years ago. In this great sweep of history the variety of a common item such as pottery is quite staggering, while the diversity of cultural repertoires increases with the transition from stone- to metal-based technologies.

It is this process of amplification that gives us the sheer variety of material goods with which we play out social lives (that are still structured by Dunbar's number). Even as far back as 300,000 years ago, as archaeologist Larry Barham has pointed out, we see an amplification in technology as stone was hafted to wooden shafts to form the first composite tools. This may pale into insignificance compared to the thousands of diverse components in an automobile or the bewildering variety of goods on sale in a superstore. But judged by the times, it represents an amplification of technology. It also warns us against dismissing the handaxe-using hominin from Herto as unchanging.

But what else might have changed? Here a second penny dropped. Social life is not just about artifacts such as grave goods and the built environment that gives it a particular form. For the social brain – and here the insights of primate watchers on the project were invaluable – social life is about interaction between dyads such as a mother and her child. At this level social life is about building bonds that last. Our emotions are one of the core resources that makes such bonds as strong as they are. Using our mentalizing skills we turned a survival feeling like fear into a complex social emotion such as greed. Here the cognitive skill of reading another's moods helped amplify our understanding of what was taking place.

Emotions can always be made stronger by amplifying them either with or without the use of technology that draws on that other core resource, materials. Music provides a good example, whether involving singing alone or with others and then by adding in instruments and building an orchestra or a band. The result changes moods and enhances social interaction. And this needs to happen because, as the numbers in a group increase, so the pressure mounts to intensify those bonds as more social partners need to be incorporated.

What we suggest as a solution to the archaeological conundrum of 'no change' is that an analogous process to the amplification of technology and culture took place among our emotions. In this chapter we discuss what the emotional bonds behind music and kinship achieve, and how religion has a role to play in our emotional and cognitive development. All of these topics can take place without the support of technology. In particular, the emotional intensity of music and religion has been amplified by harnessing the senses, which underpin them, to new forms of technology such as drums, incense burners, dance halls and the echo of a choir around a fan vault.

Well, the penny might have dropped for us but we suspect most archaeologists will be sceptical about our solution to the conundrum. To be convinced, they will still want to see the material trace of a heightened emotion because that is the way archaeology works as a historical rather than experimental science. But the problem is that such a minimalist approach leaves so much of what it is to be a hominin out of the story, returning us, sadly, to the stereotype of the witless caveman surviving by raw instinct in a world that is barely social.

Most archaeologists we know examine the weight of evidence and come to the conclusion that a 'true' humanity emerged only very recently, through 'revolutions'. But theirs is an approach dictated by the principle of WYSWTW (What You See is What There Was). If WYSWTW is true, then their approach would be the right one. But if we know that most evidence decays rapidly, and the older it is the more is gone, then we have to take that into account. There is undoubtedly more to humanity than the bits of chipped stone, butchered animal bone and simple shell necklaces that survive. We have to find a framework that has space for all the other hallmark human traits – kinship, laughter, language, use of symbols as well as music and ceremony. The social brain with its comparative, interdisciplinary breadth, is one such framework. And moreover one that can aim to see how far back these traits appeared deep in our Palaeolithic ancestry.

Much attention has focused on the 'when' of this long-term process, but far less on the 'how'. In this chapter we set out to remedy this picture. The social brain allows us a different approach from most researchers, because it lets us bring right up the list those ever-present core resources – the senses and materials – from which hominins have always fashioned their immediate, and all-important, social bonds (see Figure 3.2). We will look at some of the novel social forms that have been created to support the business of being social. These include the appearance of social emotions, such as compassion, and music, which enhanced ceremonies that would come to include concepts of afterlives. These played on the senses, amplifying them in new ways to create stronger bonds that could encompass larger numbers and, on occasion, greater physical separation. But at the same time materials were available for strengthening those all-important bonds. We have already noted (in Chapter 5) the significance of composite tools. These became common in the hands of large-brained hominins such as *H. heidelbergensis* and Neanderthals, and came to dominate the technologies of *Homo sapiens*. They incorporated many materials and sets of components, like the multiple stone blades to form the blade of a knife. This interest in composing things, bringing foreign elements together to make new tools, is an indication of people constructing and living in a more complex world, extending in the end to signs and symbols. Consequently, notation, figurative art, ornamental display and colouring using materials such as ochre became widespread. The end result was that humans themselves became composite artifacts: dressed in clothes, encased in armour, festooned with jewelry, coiffured, perfumed, tattooed and painted; and such composite creations were constantly amplified further to produce cultural diversity and individual variety.

Large-brained *Homo*

To weigh up the arrival of these factors, we need to look more closely at the three hominins introduced at the start of this chapter. By 500,000 years ago the threshold for large brains, which we set at 900 cc, had been well and truly crossed. *Homo erectus* (see box on p. 120) just manages to creep into the lower end. But it is the three closely related descendant species – *heidelbergensis*, *neanderthalensis* and *sapiens* – that are the true large-brained hominins (see Table 6.1).

A few finds across the Old World suggest that the new human form, *Homo heidelbergensis*, emerged as early as 800,000 years ago. Palaeoanthropologist Philip Rightmire has discussed this phenomenon in a series of papers. It is striking how few specimens are available for making comparisons – three or four in Africa, and a similar number in Europe. In Africa, only the Bodo cranium from Ethiopia is well dated, to about 600,000 years ago, while the ages of Kabwe and Elandsfontein further south in Africa remain uncertain. But Bodo, at 1250 cc, is sufficient to show that in at least some cases brain size in African human specimens had risen some 200 cc beyond that of *Homo erectus*, a rise of around 20 per cent that takes us into the modern range. The crania of Ndutu (near Olduvai) and Kabwe (in Zambia) are massive, but with higher vaults than *Homo erectus*, and more modern features of the face and palate. Rightmire believes that the new species can be called *Homo heidelbergensis*, because of the similarities with later European specimens. The latter are all later in date, emphasizing the sparse nature of the primary data. Even so, there is no doubt that somewhere early in this new era, the Middle Pleistocene (*c.* 800,000 to 125,000 years ago), the long stasis of *Homo erectus* was broken and the evolutionary direction led to larger brains.

Among the vast areas lacking in preserved hominin remains, these few crania make a good ancestor for all later humans who have survived. To survey these we have to turn to ourselves, modern *Homo sapiens*, as well as to our cousins the Neanderthals, *Homo neanderthalensis*. That still leaves other large parts of the world, and just the thin scatter of finds. But we are enormously aided by our new knowledge of whole genomes. These show beyond doubt that *Homo sapiens* and *Homo neanderthalensis*

	Brain size, average cc	Theoretically predicted group size	Age range '000 years
Homo sapiens (Before 11,000 years ago, worldwide)	1478	144	200 to 11
Homo neanderthalensis (Eurasia)	1426	120	300 to 30
Homo heidelbergensis (Europe: Steinhem, Petralona, Atapuerca)	1240	128	600 to 250
Homo heidelbergensis (Africa: Bobo, Kabwe, Ndutu)	1210	126	800 to 250

Table 6.1: *Three large-brained hominins. The smaller size than might have been predicted for Neanderthal communities takes into account the configuration of their large brains, as explained in the box on pp. 158–59.*

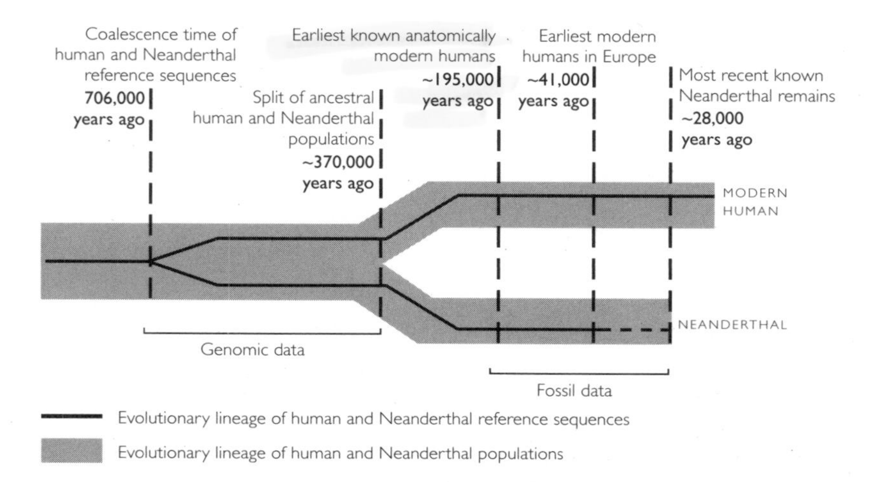

Figure 6.1: *The branching of human populations through the last million years as shown by genetic and fossil data. The dates are necessarily approximate. Recent research suggests that Neanderthals and Denisovans (not shown here) are sister groups whose ancestors separated about half a million years ago.*

came from a common stem, and suggest that it existed around half a million years ago, a perfect fit for *Homo heidelbergensis* (see Figure 6.1).

Similar but different

What does a large brain mean in terms of personal network size? Remember, we have seen that the appearance of *Homo* – the first members of our genus – around 2 million years ago is marked by a distinct upswing in the size of communities to something between 80 and 100 individuals (it increases slightly with time). However, community size remains broadly stable for the better part of 1.5 million years.

Then, around 800,000 years ago the oldest of the large-brained hominins, *Homo heidelbergensis*, emerged from the African *Homo erectus* populations. We argue that this led to a steady increase in brain size and community size through time, one that was crucial in the development of a more organized social life. The populations of *Homo heidelbergensis* in the north eventually gave rise to the Neanderthals (*Homo neanderthalensis*), with their distinctive anatomical adaptations to life in cold climates at high latitudes. The Neanderthals were in fact an incredibly successful species, occupying Europe and western Asia as far east as the edges of Siberia from around 200,000 years ago or earlier until they disappeared sometime after 30,000 years ago. With their heavy, stocky bodies, they were able to cope with the problems of heat loss that high latitude species inevitably face, especially in winter. And the greater strength their bodies gave them allowed them to develop a style of confrontational hunting that proved very successful with the large herds of deer, bison and mammoths that they found on the plains of Europe south of the northern ice sheets. Using heavy thrusting spears, they were able to take on large meat-rich prey species, and their protein-rich diet in turn allowed them to build heavily muscled bodies.

And here lies a question for the social brain. If Neanderthals and *Homo sapiens* had such similar-sized brains, how was it that they differed so much when it came to technological skills? Our answer may be that for a long time they did not – until about 100,000 years ago the Neanderthals showed all the skills and practices of their southern cousins. In some way, however, the Neanderthals seem circumscribed or bounded. They occupied their northern territories across a wide swathe of Eurasia, but what they never did was to leave this homeland and settle unfamiliar lands either at lower latitudes or above the Arctic Circle. The latter would have opened up the possibility of moving further east towards the Bering land-bridge* and so eventually into the Americas. In a striking contrast, these opportunities were eventually taken up by *Homo sapiens* in their global dispersal that began from Africa sometime after 100,000 years ago. One solution to the problem of why Neanderthals and humans are so different is presented in the box overleaf.

From hominin to human: some shared factors

Big brains were necessary for the change from hominin to human, but they are not a sufficient explanation, any more than is technology. The building of social worlds is closer to the heart of things, and in this several other factors were very important – we will single out here especially music, kinship and religion.

Music and emotion

We saw in the previous chapter how something as simple as laughter could increase the size of the grooming group. We have also argued that language

* At periods of low sea levels, such as during the Ice Age, when water was locked up in the ice, a so-called land-bridge was exposed in the Bering Strait between Siberia and Alaska, linking Asia with the Americas.

was co-opted earlier in the hominin story to solve grooming problems within larger communities. Indeed we suggested that language grew out of laughter. But at that point, we hit a glass ceiling again. Something else was needed to allow our three large-brained hominins, *H. heidelbergensis*, Neanderthals and *H. sapiens*, to lift community size still further.

That something was music – or, to be more precise, singing and dancing – as Steven Mithen has also recognized in his book *The Singing Neanderthals* (2005). These aspects of music provided a transition from laughter to speech. The singing would have been a purely wordless form of chorusing, perhaps derived from communal laughter. The key to it is that the activity is very rhythmic and hard work for the muscles. Singing is much more demanding than speaking, and dancing is of course very physical. Music-making in its various forms thus also turns out to be a good trigger for endorphin release, as we saw in Chapter 2. Our experiments demonstrated that it is the physical work involved in music-making that is important: passively listening to music does not produce the same effect.

Music-making has an important additional property, namely that it allows a larger group to be involved than is the case for laughter. We don't know what the upper limit on singing and dancing groups actually is, but it is certainly larger than the three or four individuals found in laughter groups. However, the significant point is that these musical activities have another property that appears to have important consequences for endorphin release: they are highly synchronized. Performing in synchrony seems to ramp up the levels of endorphin release compared to doing any of these activities individually, a perfect example of amplifying what already exists to provide a solution to a social problem. It also explains why we like to march in step, pull together and sing in unison.

Kinship and mentalizing

Among humans kinship is a very powerful bonding force. It is a good example of social storage where obligations, connections and emotional ties are neatly put in the box marked 'family'. Favours can be called on without thinking and with little expenditure of social capital. To ask the same of friends is a different proposition.

The kinship of social anthropologists can be difficult to trace among archaeological data. But from an evolutionary timescale some signposts

Neanderthal big eyes and large brains

Why are the Neanderthals so different from modern humans, when both descended from a common ancestor of about half a million years ago? One reason may be that for long periods Neanderthals lived much further north than all other hominin species. Living at high latitudes incurs one particular problem that is not faced by species that live in the tropics: low light levels, especially in winter. There are two aspects to this. First, as you move away from the equator towards the two poles, light from the sun cuts through a progressively deeper layer of air due to the earth's globular shape. This naturally reduces the strength of the sunlight. And this is further reduced by the usually greater cloud cover at high latitudes, especially in the northern hemisphere. Secondly, on top of all this, there is a much greater seasonality in the climate nearer the poles, thanks to the fact that the earth's axis to the sun tilts progressively over the year, with the sun being over the southern tropic during our northern winter and over the northern tropic during our summer. The result is the dramatic variation in day-length across the year that gives us our seasons, with short days in winter and long days in summer. Species that live at high latitudes have to be able to cope with long dark nights and short dull days.

A common evolutionary solution is to increase the size of the retina (the light sensitive layer at the back of the eyeball) so as to absorb more of the light that is available. (Essentially the same principle is used in night binoculars, which enable more light to be gathered in and focused through the pupil of the eye onto the retina at the back.) And if you want a bigger retina, then, as a matter of simple physics, you must have a bigger eyeball to house it. Now, Neanderthals had unusually large eyes (about 20 per cent bigger than those of modern humans). They were also characterized by a distinctive feature: the so-called Neanderthal 'bun' at the back of the head. Our skulls are rather globular in shape, but theirs were more elongated, with this bulge at the back. The back of the brain is the area that is dedicated to visual processing, and across primates as a whole the volume of the visual areas in the brain, the volume of the optic nerve and the various way stations in between, and the volume of the eyeball all correlate very tightly together. This is not too surprising, because the visual system as a whole is organized like a very simple map: each area in the retina maps onto matched areas in the successive layers of the visual cortex. So if you have a bigger retina, you must have a bigger visual system in the brain to process the incoming light signals. Therefore if Neanderthals had bigger eyeballs, it follows that they had bigger retinas, and bigger visual areas in the brain. And as a result, they would have had bigger overall brains to accommodate these – unless, of course, they were prepared to sacrifice some other part of the brain for better vision.

For Neanderthals this may have been the case. Unlike their sister-species, *Homo sapiens*, they seem to have been conservative in terms of brain evolution. In other words, even though their total brain volume increased, the 'business' (or frontal) end of their brains remained unchanged: it was only the visual system areas at the back of the brain that increased in size. As a result, their community sizes will have remained much the same as those of the archaic humans from whom they descended. In contrast, something caused a real functional increase in brain size to occur in those populations of their African cousins that gave rise to anatomically modern humans (i.e. us). Because these were evolving in the tropics, they did not need bigger visual systems, so all the increase was at the front of their brains. And since the frontal lobes seem to be crucial in determining social group size, this allowed them to increase community size from the 100 to 120 typical of archaic humans and Neanderthals to the 150 that now characterizes us. When these populations later invaded Eurasia (around 70,000 years ago), they too would adjust brain size to cope with lower light levels – but by then they had already increased the size of their social brain, and so they added extra visual system volume on rather than sacrificing frontal or temporal lobe volume.

In order to test whether the Neanderthal bun was indeed part of an adaptation to solving the problem of low light levels at high latitudes, Ellie Pearce, one of the postgraduate students on the Lucy project, looked first at modern humans from different latitudes. She measured the brains and eyeball sockets (or orbits) of skulls in museum collections, and correlated these with the latitudes where the owners of these skulls had lived. These were, by the way, all people who had lived within the last 200 years. She found a strong relationship between orbit size and latitude, and a similar correlation between cranial volume (more or less equivalent to brain size) and latitude. In other words, people who come from higher latitudes actually do have bigger eyeballs and bigger brains to match. However, Pearce was also able to show from a different dataset that visual acuity (the ability to discriminate shapes) hardly varied at all with latitude. In other words, as modern humans have migrated further north (and south) from the equator, they have increased the size of their visual systems to compensate for the lower light levels so as to keep visual acuity roughly constant. It seems that the Neanderthals inhabited northern regions critically too early to be able to develop the full social brains that characterize *Homo sapiens*.

 can be seen. Anthropologist Alan Barnard has described a universal kinship that allows modern hunters and gatherers to move freely across large areas. Kin in this instance goes beyond a close genetic relationship. It is more a case of calling a complete stranger 'aunt' or 'uncle' in order to make a bond using the idiom of kinship. Barnard believes that universal kinship was the earliest form of kinship and underpinned all human societies in the last 200,000 years, including those of anatomically modern humans. He may well be right. But equally it could have appeared as a development from those bonds based solely on the closeness of two people's genes. In other words, the great push by modern humans that started 50,000 years ago beyond the bounds of the Old World to Australia and then the Americas was helped by this development; a change that made it possible to stretch social relations across space and time. This was liberating in the sense that absence from social partners no longer severed social bonds. Instead the bonds were amplified by recasting them according to the cultural rules of kinship. We described in Chapter 2 how students going to university changed their friends but not their family. The strength of kinship overcomes distance and the time needed daily to build and maintain relationships. Those freshmen are, in microcosm, the equivalents of the first settlers to reach Australia 50,000 years ago, and for that matter any new environment where distance and time stretches the ability to stay in touch to the limit. The solution has always been to forget friends and trust in kin.

The creation of distant relations through kinship also opened up the opportunity to bring in the dead. The ability to 'go-beyond', which at the world scale saw humans settle new continents, also allowed them to go in another, imaginative direction – the worlds of the afterlife and the ancestors. The theory of mind required to accomplish this task is considerable. The orders of intentionality, or steps in social reasoning, become hugely complex as the motives of imagined people and beings, that now include gods and spirits, have to be factored in.

'The penny dropped and Henry *believed* his fairy godmother who *wanted* him to marry Jane and so fulfil the wishes of his dead uncle rather than *second-guessing* the call of her heart which *whispered* that Roger *intended* to propose and so redeem his family's honour.' Such fifth order intentionality stretches our imaginations as well as our social reasoning. It is like seeing the right chess move five steps ahead. Kinship makes some of this mentalizing simpler, because people are not behaving as independent individuals but according to roles defined by what sons, aunts, grandfathers and cousins are expected to do. Fairy godmothers and dead uncles are also part of that complex universal kinship, and their presence points to very high orders of intentionality.

We saw in Chapter 2 that mentalizing abilities are correlated with brain size (and specifically frontal lobe volume). As a result we can estimate the mentalizing competences of different fossil hominin species, and importantly, do so with reasonable confidence in the accuracy of the estimates. These estimates suggest that *H. heidelbergensis* and Neanderthals would have been able to manage fourth order intentionality. This would have had a very significant limiting effect on the grammatical complexity of their language – and on the complexity of the stories they could tell.

By contrast the data suggest that fifth order intentionality was achieved by anatomically modern humans around 200,000 years ago at the earliest. This implies that a *Homo sapiens* fossil such as Herto, with a simple stone technology, nonetheless had superior mentalizing abilities. This may seem surprising until we remember that two of the skulls from Herto were deliberately modified post-mortem. The cutmarks on the crania and jaws show they had been intentionally disarticulated and defleshed. They had also been scraped and polished. The great African archaeologist, J. Desmond Clark, described this as decoration. There are also cutmarks but no decoration on the older *H. heidelbergensis* skull from Bodo. Now these post-mortem rituals may not match the mentalizing example of Henry and his fairy godmother, but they do provide a material trace of the more complex stories that large-brained hominins were capable of. The archaeological mantra WYSWTW is challenged by the Bodo–Herto evidence, which itself is predicted by the social brain.

Religion and storytelling

The presence in large-brained hominins of laughter, music and then full language allowed a third endorphin mechanism to kick in: the rituals of religion. Many sacred customs such as fasting and penance (not to mention martyrdom) are well designed to put stress on the body, and so generate those crucial endorphin surges. But knowing that you should do these rituals, and why, requires language. Some 50,000 years ago religion would have been shamanic, a form we still find among hunter-gatherer peoples and other small-scale societies. Its features were memorably described by David Lewis-Williams in his book *The Mind in the Cave* (2004), where he drew parallels between the shamans of the Kalahari and the therianthropic creatures (part-human, part-animal) depicted in the art of Upper Palaeolithic Europe.

Shamanism is a religion of experience rather than a religion based on doctrine. These kinds of religion do not normally have a theology, nor do they involve belief in gods of a conventionally understood kind. Instead, they use music and dance to create trance states during which the adept enters a spirit world peopled by therianthropic animals and ancestors, some of whom are dangerous and some of whom act as benign spirit guides. But they need language, and perhaps a sophisticated language ability, to be able to describe the travels in the spirit world experienced during a trance. These kinds of religious ceremonies are also very therapeutic – almost certainly because they involve dancing and the endorphin surge that produces. Trance dances of this kind are often used explicitly to cleanse the community of bad feeling generated by the inevitable social stresses of group life.

Mentalizing also plays an important direct role in this part of the story because religion depends on being able to imagine that there is another spirit world that exists in parallel to the everyday world of experience. That minimally requires formal theory of mind. However, there is nothing to be gained from just imagining such a world: to make it

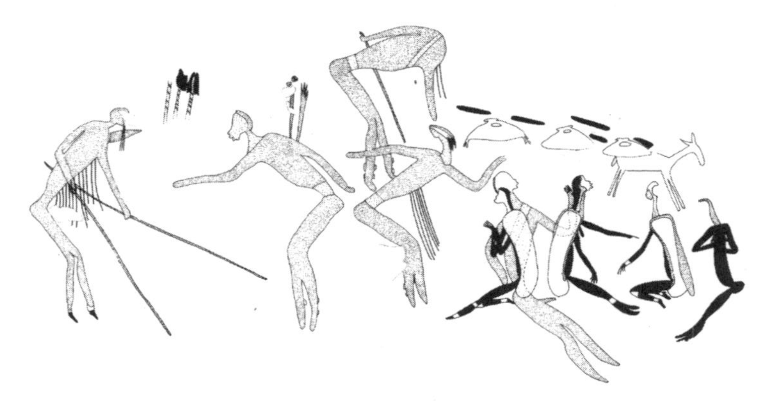

Figure 6.2: *The figures of South African rock art provide a basis for the ideas about shamanism developed by David Lewis-Williams and colleagues. The image shows the shaman's imaginary journey to the spirit world. A journey only made possible by an advanced theory of mind.*

a religion, you have to be able to talk to others about it, and minimally that requires you, me and a third party, as well as one or more minds in the spirit world (an ancestor or even a god). That gives us fourth order intentionality as a minimum requirement for religion and might well mean that large-brained hominins such as Neanderthals had beliefs about a spirit world, but the extra order of intentionality that modern humans have allows those beliefs to be significantly more sophisticated.

The difference between fourth and fifth order intentionality is much easier to appreciate in the context of another important mechanism involved in social cohesion in large communities, storytelling. Having a shared world-view involves being able to exchange views and opinions about how the world is or might be, and stories are often the medium for doing this in traditional societies. Origin stories remind us who we are and how and why we came to be, while folk tales are often vehicles for evaluating moral dilemmas. The quality of story that we can tell (and understand as the audience) depends on the number of orders of intentionality that we can manage – not least because these determine how complex our sentences can be. The difference between fourth and fifth level intentionality in terms of the quality of the stories is quite striking. So, while archaic humans and Neanderthals, with their fourth order intentional competences, could certainly have told each other stories, the quality of those stories would not have been in the same league as those that anatomically modern humans could have told. In short, modern culture as we know it is unlikely to have evolved before the appearance of anatomically modern humans, *Homo sapiens*, around 200,000 years ago. And even then it took a long time for all the components to develop.

Big brains, but what was happening?

Three big-brained hominins – but only one, *Homo sapiens*, seems to turn up the volume. But for a long time *Homo sapiens*, as an anatomically modern human in Africa, seems to have kept the volume low. At least that's what those many archaeologists who support a recent 'Human Revolution' would argue. Their views were strongly articulated in symposia organized by Paul Mellars, Chris Stringer and Katie Boyle, published in 1989 and 2007, and are expressed in similar vein by leading archaeologists such as Ofer Bar-Yosef and Richard Klein.

The idea of a recent (and by recent they mean after 50,000 years ago) Human Revolution is driven by the WYSWTW approach. However, strict adherence led its most fervent supporters into trouble, as Sally McBrearty and Alison Brooks pointed out in their aptly titled paper 'The Revolution that Wasn't' (2000). Various notable items of material culture, including pieces of ochre, engraved or plain, shell-bead necklaces, long-distance trade in raw materials and novel subsistence items such as shellfish – all supposed items of the revolutionary package – turn out to have a long ancestry in Africa prior to being found outside the continent (see Figure 6.3). Rather than a revolution, there was instead a prolonged rise in the capabilities of anatomically modern humans, one that eventually outdistanced the performances of other communities such as the Neanderthals and no doubt other, as yet undiscovered, regional populations in Africa.

This is where the social brain comes into its own as a new type of enquiry. Rather than being dictated to by a WYSWTW approach, we can step back and ask what was important for the longer-term change from hominin to human. We examine this by asking two questions:

- What did change when nothing seems to change (harking back to our alien visitors of 500,000 years ago, when they would have found that large brains were not matched by a corresponding amplification in new tools and ways of doing things)? This is a question about how the use of the senses became amplified to heighten social interaction.
- What social changes are needed to explain the massive developments in global settlement and population numbers that happened in due course? These began well before agriculture and when societies were still small in scale and based on extracting rather than producing food. This is a question about amplifying the social signals that artifacts contain to ensure successful social interaction took place.

What underlies both questions, and indeed our whole approach to the archaeology of the social brain, is that the materials and the senses that form the resources for making social bonds *do not have to evolve together*. They are not a tandem-bicycle requiring both cyclists to put in

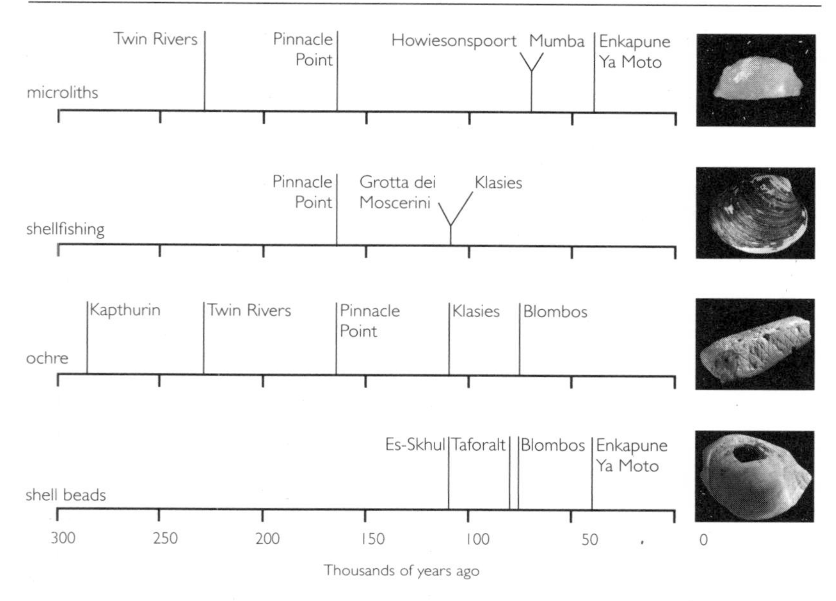

Figure 6.3: *The timelines of four key lines of evidence pointing towards higher levels for sophistication in human behaviour – microliths, shellfishing, ochre use and personal decoration with beads.*

the same effort. Our hypothesis is that the senses and emotions can be amplified independently of materials and artifacts such as stone tools. And the reverse is also expected. In the tandem-bicycle analogy one cyclist does the work while the other freewheels for a bit. However, they are so closely linked by virtue of being the resources that are the building blocks of any hominin and human society, that changes in one can precipitate change in the other. They are engaged in a co-evolutionary process, but the contributions, like those of the two cyclists on the tandem, are not always equal.

Channelling emotions

Homo erectus, as we saw in Chapter 5, reliably did the same things again and again and again. It is as if they were like us, but missing some vital dimension, of creativity, or insight. Or was it more a matter of temperament? This is an interesting thought, because the apes – at least the chimpanzees – are actually far more emotional than us. Anything that happens in their community is the cause of great excitement. Some researchers point out the importance of homology, or shared ancestry – the primary emotions go back far beyond our own genus, and have enough in common that we can recognize them in a dog as much as a chimpanzee. We share not just with them but with every mammal fear, anger and sexual drive, and like them we are driven to action by hunger and thirst.

There are important contrasts too, that begin to leave evidence in the record. Philosophers and scientists – including Darwin – have picked out the human capability of 'rational' thought. That leads to the belief that rational thinking is the human ideal, to be contrasted with a more primitive and less desirable emotional approach to life. Yet there is a powerful school of thought that distinctively human emotions offer much to shape our nature.

The behavioural scientist Dylan Evans has expressed this value of the emotions in the 'search hypothesis', which sees the emotions as there to set limits to our actions, to provide effective spur-of-the-moment judgments: 'Looking before you leap is all very well, but the point is to leap. At some point you must stop thinking and start acting.' He calls this Hamlet's dilemma. Others also see the emotions as deep-seated 'rule of thumb' guides to leading life. You might need to decide quickly whether to go out on Thursday night to a football match, for a meal with friends, or to stay at home. The feeling 'I'd like this, but I would not like that' will often rule over a strictly rational analysis. It can be argued that we are guided by a whole array of deep-seated 'dispositions' or emotional prompts, which the anthropologist Pascal Boyer describes as 'biological systems'. They are undoubtedly systems, in the sense that if we are out in the bush, and we suddenly sight a lion, we will feel an immediate visceral shock; if we then realize it is only an antelope, we will feel an immediate counter-reaction – and in these circumstances a physiologist could detail the sequence of reactions that we went through.

What makes modern humans stand out is a unique and immensely powerful ability to manipulate and stimulate the emotions through choice. The necessary connections are not in the old brain, the limbic system, but wired through the neocortex, which we have already mentioned a number of times in the context of the social brain. We can choose to read Shakespeare's *Romeo and Juliet*, so as to make ourselves sad. We can contrast fictional tragedy in our minds with real tragedy. Or we can terrify ourselves with Corbett's *Man-Eaters of Kumaon*, the supernatural horrors of Edgar Allan Poe or a no-holds-barred splatter film. Or – a much more real danger – we can choose to think aggressive thoughts, and even place them in the minds of others. Is that not usually the role of a demagogue, from the time of Demosthenes and Cleon, or perhaps long before?

Did *Homo heidelbergensis* perhaps daydream by the fireside? At Beeches Pit in Suffolk, where John Gowlett led excavations alongside many other researchers, early people sat around their fires by the side of a pond or creek. As he remembers the excavation, 'We found the butt of a handaxe. I hoped we might find its other half, and eventually spotted it among other emerging finds. I let the students excavate it and recognize it themselves. It was just half a metre from the butt. Probably any modern human would have reacted to breaking a favourite tool – you or I might have thrown it into the pond.' But *Homo heidelbergensis* placidly left the two pieces side by side.

Making special

Vital as they are, emotions are a muddle – ours are greatly dampened compared with those of chimpanzees and bonobos, but gorillas can rarely be accused of over-excitement. Rather our emotions have been reconfigured, and at some point in the past they did light up. They leave us as creatures who laugh and cry, find music runs deep into our souls, and who can communicate through poetry. We would be pretending if we believed that archaeology traces the origins of all such things, but we do have reason to believe that these major changes began to take shape through the last 500,000 years of large hominin brains – and the social brain does offer a framework that allows us to test our fragments of data. Here we concentrate on the best evidence that we have – the burials, the beads, and a new technological complexity. Art would be our further great aid, but its pattern of preservation is exceedingly patchy. We will come to art, but first we can extract an important idea, that part of art is simply 'making special', and that we can begin to find this making special on a much broader basis.

Another aspect of making special is 'added value'. By this we mean that things can acquire additional meanings, sometimes more important than the original function. We see the first hints of this 400,000 years ago, when a fossil shell embedded in a handaxe might perhaps be the dominant feature of the tool rather than its symmetrical shape or sharp cutting edge.

We have to accept at once that the beginnings of these new features are exceptionally difficult to pick out, and to date. That is why archaeology cannot tell the full story on its own. Even so, we can note some basic but telling developments in technology. There were shifts in the dominant form that artifacts took. Since 2.5 million years ago tools had been dominated by the hand and the instruments it could wield. Now, during this later 500,000 years, people experimented with new ways of organizing their material worlds. One approach was unleashed by the use of fire. Neanderthals gained such control of burning temperatures that they could make pitch, which requires sustained heat over a long period. Made from pine or birch resin, this was an effective glue, giving testimony to a new degree of combination. Another route led to a conceptualizing of the world based on ideas of containment. The ideas may well have started with items as basic as bags and buckets (which have not survived). However, the main archaeological evidence for this development comes in the form of huts and houses. The evidence is sometimes contentious, but large-brained hominins like Neanderthals were building roofed shelters in their rock home at Abric Romaní in Spain – an artificial home within a natural home. These are amazingly preserved in the form of natural casts, skilfully excavated by the team of Eugene Carbonell.

Ceremonial hominins

Anthropologist Wendy James has characterized us as the 'ceremonial animal'. We operate in prescribed ways, and need to do so. Deficiencies that can easily arise in the system actually highlight the power of its operation. A British Member of Parliament admits to showing a syndrome of obsessive compulsive disorder, of needing to do things in fours – on entering a room he must switch the lights on and off four times. Less extreme tendencies to detail affect most people: if you sit at a table with a chequered cloth, you and your companions may well unconsciously (or consciously) move glasses and cups to the 'fitting' places on the design.

All this is perhaps less surprising since we are the descendants of *Homo erectus*, who a million years ago so much needed to make a tool 'like this' rather than 'like that', that the narrow sets of ideas survived for hundreds of thousands of years. The essential thing again is that we 'make special'.

For Wendy James, 'Ritual, symbol, and ceremony are not simply present or absent in the things we do; they are built into human action.'

The beginnings of 'making special' first govern ordinary routines, then beyond the ordinary it seems it begins to allow and need humans to pick out the extra-special. Why should that be? Very likely, it is that larger societies require stronger signals.

It is noticeable, for example, that in most modern armies only the most senior officers wear red tabs or flashes, at the level of 1 staff officer to 500 or 1000 troops (or more). What these flashes do is signal attention. Humans are great at concentrating. We can lose ourselves in a crossword puzzle or a good book or anything that grabs our attention. Chimpanzees do not have that ability. They may spend minutes grooming each other with a look of absorption, but such rapt attention is never transferred to making or using things (except perhaps in the tasks of feeding). Two aesthetic qualities capture our attention and so are ideal for this kind of signalling, and the evidence shows that our large-brained ancestors were equally interested in them – colour and glitter. They allow projection of specialness at this very high level of distinctiveness that is not usually necessary in the smaller society.

For us, gold, silver and diamonds represent the extreme case, though they were beyond the extractive abilities of earlier humans. Throughout this book we have argued that the technological is also social. Modern cases blur lines. If someone gives you a birthday cake, that is clearly a social act. If the cake is good to eat, it touches on subsistence, but it is also a work of technology. What the cake does especially is to put emotional value into the outside world. We modern humans go overboard with the value, if not the emotion, when it comes to the precious metals and diamonds. They can 'mop up' enormous value, equating also to being very hard to get (we might contrast rare earths, immensely valuable in money, but with no intrinsic attraction). Surely such a profound addiction must have deep roots in our evolution. But gold, silver and diamonds are all recent interests, products of technologies less than 5000 years old. What were their predecessors? In a few words they seem to have been shells, red ochre and volcanic glass.

The first hint of an importance in shells may come with the presence of fossil shells as apparent centrepieces in the middle of one or two handaxes made 400,000 years ago and first noted by Kenneth Oakley.

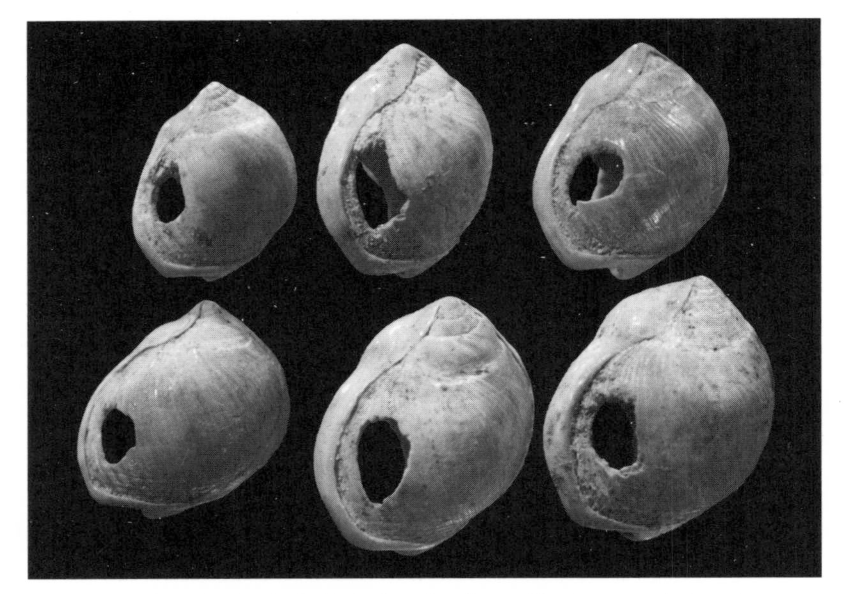

Figure 6.4: *Pierced shells, such as these found at Blombos Cave in South Africa, would have been strung as beads – as the wear evidence confirms – thus proving both the existence of string or cord and the importance of personal decoration.*

We could argue that their presence is accidental. But they are an inconvenience to the process of stone knapping, and it would probably take a deliberate sequence of actions to keep them at the centre of the piece. By 120,000 years ago, certainly, shells are being pierced and strung, probably in necklaces. At Blombos Cave in South Africa, a 70,000-year-old set of shells was found together, demonstrating this point; and at a similar date the same species of shell, perforated in a comparable manner, has been found in excavations at the Grotte des Pigeons at Taforalt in Morocco.

In material culture colour is very important to modern humans. Red, indeed, may be the trigger of one of Pascal Boyer's biological systems, a sign of danger. In human activity it appears first as ochre on archaeological sites, as much as a million years ago, at Bizat Ruhama in Israel. Red ochre is widely used in modern times and in later prehistory, striking as a pigment used in art. It features in dots in the earliest dated cave painting. It can be taken to symbolize life, or blood, or more complex ideas. We should not forget that it can also have practical uses, for preparation of hides, lubrication or as a constituent of glues. Through the last 300,000 years in Africa in particular, ochre became so important that it was mined in quantities, as at Twin Rivers in Zambia. At the Kapthurin site in central Kenya a collection of about 5 kg (11 lb) of ochre was found, dated to about 300,000 years ago.

At Sai Island, in the Nile in Sudan, polished and ground ochre has been found dating to about 200,000 years ago. Our colleague Larry Barham stresses that the finds have their roots back in the Acheulean. The archaeologist Lynn Wadley and others again have stressed that there are industrial uses as well as 'symbolic' ones, but there is no doubt that added value begins to adhere to this material somewhere in this period. The symbolic importance of ochre is apparent at Blombos Cave on the South African coast. Here, dating to about 70,000 years ago, is a piece of engraved ochre with a lattice design carefully scratched into the surface.

Stones could also have a very appealing appearance. By now people had been selecting them for more than 2 million years. The volcanic glass obsidian is one attractive material, occurring in blacks, browns and even reds. It is very familiar to us in the central Rift Valley of Kenya where John Gowlett works. On occasion it would be used in early times for handaxes. At Kariandusi, first explored by Louis Leakey, such handaxes are common. Some 50 km (30 miles) away at Kilombe, it occurs very

Figure 6.5: *A small block of red ochre from Blombos Cave in South Africa bears the oldest known geometric engraving. It is often taken as evidence for symbolic thought. Ochre is a natural pigment widely used to change the colour of bodies and things as well as being the medium for much rock art.*

rarely among all the lava handaxes. Even a million years ago this material was attractive enough to transport on occasion, but the real change came in the last 200,000 years. In these later times obsidian was coming to Kilombe from the Naivasha region more than 100 km (60 miles) to the south, used for hundreds and hundreds of small flake tools. Lucy project research student Dora Moutsiou has studied the increasing use of this material across the world after 40,000 years ago. She has shown also how it was moved much greater distances from its volcanic source when associated with modern humans, often by a factor of two. Crystal-clear quartz also has an appeal to both modern and it seems ancient eyes, and was carried long distances in a similar way.

Our primary evidence that such materials were appreciated is seen in their transport, and in their selective use in tools. From early times, humans had to make judgments about value. There is a high cost to carrying, and selection became important, so that good things were transported, and rubbish was not. Again we see technical and social bound together in these judgments.

Burial rites

The last of the first great ritual innovations is the appearance of burial, and it has special messages for us. In the social brain we see again and again the importance of relationships, but some are especially important – those of kith and kin.

The earliest Palaeolithic burials known are found in caves, from at least 130,000 years ago in the Middle East. They include both early modern humans, at Qafzeh and Es-Skhul, and a Neanderthal, at Et-Tabun. From then on Palaeolithic burials are known widely across Europe and Asia. Classically, burial tells us of religion, people with a sense of an afterlife, of sending a relative on a journey. Some revisionists have rather spoilt this emotionally satisfying picture – the pollen of flowers supposedly placed with Neanderthal burials at Shanidar in Iraq may well be intrusive. But such revisionism is overdone. We still struggle to appreciate what kind of treasure trove we have in the burials, but it is a very substantial one. Let's work backwards: we have millions of burials in recent times; many thousands from the Classical world; and in the Levant more than 400 from the so-called Natufian period alone, about 10,000 years ago. Before this the Upper Palaeolithic has hundreds more in total, on occasion buried in groups, and with conspicuous grave goods. How strange it would be if the several hundred Neanderthal and early modern human antecedents of later burials reflected some other totally different phenomenon. Even to argue that seems obtuse, and in fact a closer inspection of the French archaeological literature (by some English-speakers in particular) would make the matter quite plain. In the cave of Qafzeh near Nazareth, the burials of early modern humans are carefully arranged in similar orientations. The same is true at Es-Skhul on Mount Carmel. The layouts show a careful respect for personal space, as the archaeologist Avraham Ronen has argued. At La Ferrassie in France, in a small cemetery in a rock shelter, the Neanderthal adults are arranged in one group, their infants nearby under low circular mounds. The infant remains are so fragile that they would have virtually no chance of surviving if they were not interred very carefully. Yet in fact they occur widely, from Western Europe across to the Middle East. In a wide-ranging survey, archaeologist Paul Pettitt has convincingly shown how these Neanderthal funerary rites marked a modernizing phase over much older funerary caching at sites such as Atapuerca in Spain, where many bodies were found at the base of a deep cave shaft. Very few burials even in this modernizing phase are found in the open, but one, at Taramsa Hill in Egypt, captures for us an early modern human, in a pit, with the legs flexed up.

Funerary practices continued to develop in Pettitt's final phase, one example being a stunning set of finds from Jebel Sahaba close to the Nile in northern Sudan. Here is a complete cemetery dating from the end of the Palaeolithic, about 12,000 years ago. More than 50 bodies

Figure 6.6: *Three Palaeolithic burials in caves. Left and centre: Neanderthal burials at Kebara, Israel, and Shanidar, Iraq; right: a modern human burial at Qafzeh, Israel.*

were buried, some in small groups. Some show evidence of wounding by flint arrowheads.

Cooperation in life and death

What do these funerary rituals mean for our social large-brained hominins? Did they have belief in an afterlife, the type of journeys that ancient Egyptians expected to take into the netherworld? We do not know and we rather suspect not. But what they did have was a concept of an after-person, the idea that the dead still had power to affect the living. This is unsurprising, since we have already allowed our large-brained ancestors an advanced theory of mind. With theory of mind a Neanderthal was aware that another Neanderthal saw the world differently from her. Moreover they were capable of social reasoning, which took the form of calculating someone else's intentions to reach a predetermined goal. The 'retention' of ancestors, through the idea that a person persists after death as a social force to be reckoned with, is entirely consistent with this mental skill.

The early preoccupation with bodies is strong, and Pascal Boyer's arguments suggest for us that it may arise from a clash of two of his 'biological systems'. One tells us of our emotional closeness to the people concerned (or perhaps distance in the case of enemies). The other reflects our revulsion against something dead or decaying, and our inclination to deal with it – we respond by both putting away and retaining at the same time.

Of one thing we can be reasonably sure: our ancestors were burying people that they cared about, very often their close relations. The grouping of burials, seen in North Africa and from Russia to France, is one example. The placing of grave goods is another. In the earliest cases, the interment of a prime joint, or some animal horns, can be doubted – it could be accidental. But in other cases the special treatment of plastering with ochre, the presence of shell necklaces, the placing of fine tools, are indisputable. It occurs up to the present day, and to suggest that the earliest cases represent something different (because, for example, Neanderthals might have been cognitively challenged) seems to us to defy the whole benefit of tracing a phenomenon back to its roots.

Figure 6.7: *Some of the most famous Neanderthal remains come from Southwest France, including these from La Chapelle-aux-Saints. Neanderthals were regarded as brutish following the description of this arthritic old man by the French palaeontologist Marcellin Boule.*

1908.37

Look at the issue another way. Neanderthals achieved fourth order intentionality. Whether they had a belief in an after-person rather than an afterlife could be debated for a very long time. But what any hominin with those mentalizing skills had were the social emotions of guilt, shame and pride – emotions that only work when a belief exists about another's belief. Compassion would be another of those social emotions so different from the primary, or survival, emotions of fear, anger and happiness which we share with all other social animals. Instances exist of old and infirm Neanderthals being cared for by the 5, 15 or even 50 tier of their community: the 'old man' of La Chapelle-aux-Saints in France, 40 and riddled with arthritis as well as other conditions, and the Shanidar 1 male from Iraq, blind and with a withered arm. And if these do not convince you as examples of compassion, then they at least speak to cooperation and greater social solidarity among those groups in the community.

In the management of fire, and the movement of materials such as stone, there is a strong element of cooperation, which frequently seems to speak about closeness in the group. Site size too tells us that people were often organized as bands (see Table 2.1), but sometimes in smaller groups and occasionally as communities.

The evidence for the larger groupings is almost exclusively associated with *Homo sapiens* rather than Neanderthals. The archaeologist Matt Grove studied the size of campsites using archaeological data from Boxgrove in southern England (*H. heidelbergensis* at 500,000 years ago) and Pincevent, a *H. sapiens* campsite in the Paris Basin dating to the end of the Ice Age. His analysis of the density and extent of archaeological materials from these well-preserved sites points to an increase, from the time of Boxgrove to that of Pincevent, in the number of people who used them, as predicted by the social brain.

In Europe some of the large campsites have a permanent feel to them. This is clearest of all in the structure of some Upper Palaeolithic huts, in Russia especially, but also in Germany and France. The Russian examples are sometimes huge, with hearths placed at intervals down the centre (see Figure 6.8). A study by the archaeologist Olga Soffer of the animal bones left behind in the pits on the sites on the Russian Plain led her to interpret many of them as semi-permanent, winter villages. Cooperation is also evident in their interest in hunting mammoth and

Figure 6.8: *Careful excavation of Palaeolithic living sites, here at Kostenki in western Russia, often provides invaluable information about animal prey and how it was processed by people operating in social groups.*

herds of reindeer. The latter dominates the prey at Pincevent while elsewhere at Gönnersdorf and Andernach along the Rhine, dated to the late Ice Age 14,000 years ago, attention focused on horses, as described by Martin Street and Elaine Turner.

Big brains: comparing Neanderthals and modern humans

We return now to the membership of the human club and what made us human. Two large-brained hominins, Neanderthals and modern humans, stemming from the same ancestor *Homo heidelbergensis*, give us an opportunity to talk through the issues and examine our own preconceptions against the model of the social brain. What lies at the heart of this comparison is the question we posed back in Chapter 1: when did hominin brains become human minds?

An unemotional genetic analysis takes nothing away from the Neanderthals, who developed into our view of a mysterious 'other' humanity. That idea beguiles, and colleagues such as Frederick Coolidge and Thomas Wynn even teach courses on Neanderthal psychology. Yet we feel the geographical separation between the two species is not clear-cut – for long ages it had permeable boundaries somewhere in the Middle East, perhaps fluctuating as climate changed, sometimes favouring one group sometimes another. Ideas, too, could sometimes pass between populations, because in principle they can flow more easily than genes. Concepts such as burials, and the widespread technique of flintworking known as Levallois, may be cases in point.

If we look at our Neanderthal cousins in more detail, we gain a fascinating picture which has been studied many, many times. Perhaps what is ringing within us is a further echo of Boyer's systems – the Neanderthals are at the same time so like us, and so different from us, and this is what creates the attraction.

We know too that they lived in the toughest of environments. True, they were in the temperate zones also in warmer periods. Perhaps for 200 or 300 generations they had it good. But for more than 1000 generations at a time they lived in deep cold, and had to adapt to its formidable challenges. Reindeer are an incontrovertible indicator of these arctic challenges at earlier times and unexpected latitudes. At La Caune de l'Arago in southern France *heidelbergensis* people were already butchering reindeer in numbers 600,000 years ago. In the last Ice Age their Neanderthal

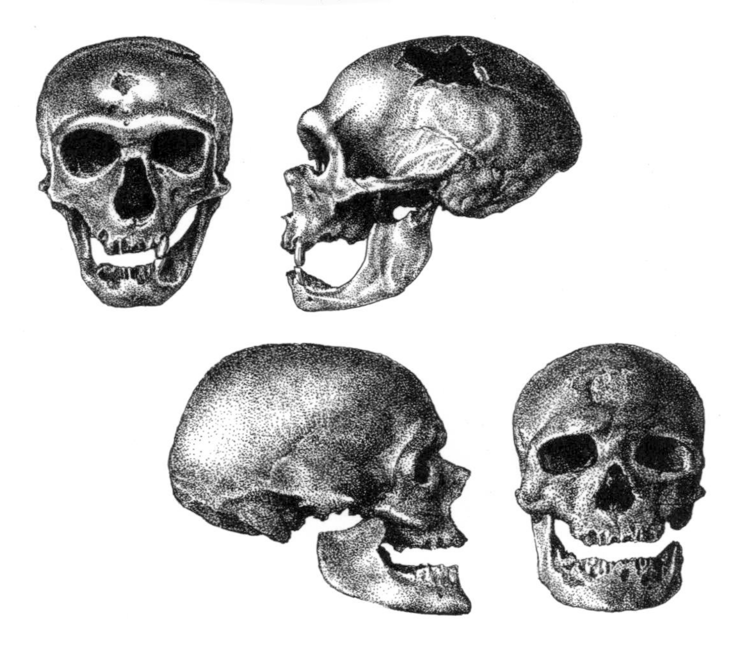

Figure 6.9: *For more than a hundred years the sharp change in the European Palaeolithic record from Neanderthals (above) to modern humans (below) has fascinated archaeologists and palaeontologists.*

descendants were systematically doing the same at sites across Europe, including Salzgitter-Lebenstedt in the east of Germany. Here Sabine Gaudzinski has reanalyzed the finds. Well able to concentrate on exploiting the single species in specialized hunting, the Neanderthals used bone points as well as flint scrapers, processing the carcasses in a way that would impress any modern hunter-gatherer. Besides their butchery skills, the hunters of Salzgitter also demonstrated the power of cooperation and teamwork. This is unsurprising. The trait of cooperation goes back for more than a million years, long before Neanderthals evolved. What they achieved, confirmed by the isotopes in their bones, which reveal what they ate, is that they were top predators.

For a long time the effectiveness of the Neanderthals as hunters was doubted. Yet in these severe climates no human could have survived without a tremendous knowledge of the natural environment and well-tried procedures for dealing with all the main demands of life.

There is still a constructive engagement between researchers who minimize the capabilities of the Neanderthals, and those who see them as standing alongside modern humans. They did perhaps live in somewhat smaller communities than modern humans, and exploited different environments. We could say the same for many populations of today, just comparing them with their ancestors of a century ago, so we must be very careful about interpreting such evidence.

An assertion often made is that Neanderthals did not have the imaginative world of modern humans, beyond the basic requirements of life. Theirs was a life shorn of the fine things: art, decoration and ceremony. We have seen, however, that there is evidence that Neanderthals cared for one another – the 'old men' of Shanidar and La Chapelle-aux-Saints who could not hunt for their food a case in point. The careful placing of other burials, and the folding of the arms as at Kebara, suggest a similar sense of care for the individual.

Compared with the Upper Palaeolithic produced by modern humans in Europe after 40,000 years ago, signs of ornament and art are indeed much rarer in Neanderthal times, almost to the point of non-existence, or at least non-preservation. Before reading too much into this, we should acknowledge that the same is often the case for later times. The late Ice Age site of Pincevent in northern France is most remarkably preserved, with extended encampments, many hearths, thousands of stone tools, and numerous bones. Even so, it preserves just one decorated bone bâton, and a smaller excavation could easily have missed this. Perhaps what is telling for the Neanderthals is what we do have: in a final phase of their times in France there is no sign that bones were being treated specially, except in one phase, the so-called Châtelperronian, found in central and southwest France. For long the Châtelperronian was thought to be associated with incoming modern humans, after about 35,000 years ago. But discoveries at Arcy-sur-Cure and Saint-Césaire indicate that the toolkits were made by Neanderthals. They include a good deal of decorated bonework as well as stone tools with a local character. We can argue, of course, that the Neanderthals would not have invented such things, but got them through contact with early modern humans living further east in Europe. This is possible, but hardly minimizes the importance of the behaviour: after all, most of us have no clue of the inner workings of the computers and phones that we use every day, let alone the ability to make them ourselves. In any case, recent finds in Iberia at Cueva Antón made by João Zilhão and colleagues, show clearly that Neanderthals were using pigments, putting them in shells. Those finds certainly pre-date any presence of early modern humans nearby. All in all the Neanderthals were certainly different – and may have thought in different ways – but their skills become notably different from those of modern humans only within the last 100,000 years.

Back to *Homo sapiens*: modern humans

Change was in the air, however, and it was from Africa that people who looked like us and shared our genes began to move. Genetic evidence gives a surprisingly consistent broad picture. Both Y chromosome and mitochondrial DNA evidence map out an ancient family tree that starts in Africa with the oldest roots, and then spreads out across the Old World. In this case it is not quite the devil that is in the detail, rather the dating is in the detail. The one fixed point comes at the end of the line, when humans first made the sea crossing to the giant continent of Sahul (Australia, New Guinea and Tasmania joined by low sea level some 50,000 years ago). It is often the specifics which we lack, along with finds from crucial parts of Asia. In consequence, those who favour a rapid but late expansion can argue that 'Out-of-Africa' started as late as 50,000–60,000 years ago, and that humans with the advantages of modernity spread ever so rapidly, reaching Australia by at least 50,000 years ago. To others the demands of that rapid journey have seemed unrealistic. Mike Petraglia of Oxford University is one who has argued for an older spread of modern humans – and new evidence of 100,000-year-old Middle Stone Age toolkits from Arabia appears to support the case.

To argue that all this is a Human Revolution needs an unshakeable belief that art and other sophisticated technologies and belief systems were the key to the great wave of newcomers sweeping around so quickly. But, as we have pointed out previously, most of the Old World had been occupied by earlier humans, much of it for a very long time. The genetics is also telling us now that the populations were moderately closely related and capable of interbreeding (most Eurasians appear to have around 1–4 per cent of Neanderthal genes). Step back a bit further, and we are reminded that large parts of any symbolic revolution may go back at least 120,000 or 130,000 years, and include the burials, the ornaments and the bone tools. Only representational art seems to be the true newcomer. And this now occurs in three corners of the Old World – Europe, South Africa and Australia – nearly 30,000 years ago, provoking at least a suspicion that it may not be such a new feature of the human endeavour.

At first glance the modern human arrivals appear to be playing the old game. People had left Africa before, so much so that archaeologist Robin Dennell has argued there was a constant two-way traffic between Africa and Asia for much of humanity's early evolution. These modern humans were also peripatetic, living by hunting animals and gathering plant foods. Their social communities were small and they lived at low densities. This was still the Stone Age.

But appearances can be deceptive. Modern humans represent a subtle change to the very framework of being human. They were more lightly built, earning them the description of gracile. Power no longer needed to be in the body and was increasingly transferred to tools such as bows and arrows, boats and traps. The social brain helps us to infer that people also had a novel social ability, the imaginative skill to project beyond the here-and-now of face-to-face contact; to participate in a society that occurred for many of its actors as much offstage as on. People were bound by what they thought others might think of their actions. Such advanced theory of mind was almost certainly present in other large-brained hominins such as the Neanderthals. But modern humans moved it to another plane. The size of their local social groups might have been only slightly larger than those of the Neanderthals, but the real difference lay in the fact that the connections and the social responsibilities became much more complicated. Instead of being restricted to a local social world, where you met most of the other members of your social network on a regular basis, these modern humans now had the capability to deal with prolonged separation and absence from their social partners. They also encountered social strangers on a more regular basis.

How do we know? One line of evidence is provided by the push from Africa across Asia and then over water to Australia. The ocean crossing, however it was achieved, implies separation. A second line of evidence also comes from interior Australia. On arrival, the first Australians rapidly settled all the major habitats of this palaeocontinent including the desert interiors which, due to the Pleistocene climates, were hyper-arid. In his excavations at the rockshelter of Puritjarra, west of Alice Springs in Central Australia, Mike Smith has found early evidence of human presence in this area. Settlement depended on reliable water sources, which still occur as rock-holes. These keystone resources were reduced when Puritjarra was first settled, implying even lower population numbers over huge areas – again indicating considerable separation between groups.

A third line of evidence comes from the raw materials used to make composite tools. Throughout the Old World we see a common pattern: after 40,000 years ago the distances increase. Stone is acquired on a regular basis from further away. Part of that would have been down to a desire for the best stone for flaking. But distance brings other benefits. The exchange of materials as gifts and later as commodities binds people together in social networks. The traded items symbolize the trading partnership and are a way of turning remote strangers into useful allies.

The geographical scale expands with modern humans. We begin to see what they were capable of as they moved through Asia into the Russian Arctic and then, with abundant landscapes before them, into the new world of North and South America. This process began slowly in the palaeocontinent of Beringia, centred on Alaska. Sometime soon after 20,000 years ago they had moved south of the massive North American ice sheets, bringing with them domestic dogs and the seeds of the tropical bottle gourd. Since these would not germinate in the cold lands through which they had to pass, we can only conjecture that they moved in the expectation of finding equable climate.

But what these seeds also point to is another major change. On the one hand we see the extension of society across space, with people literally stretched by chains of social connection symbolized by artifacts such as pearl shell, amber and figurines, while on the other hand we see the start of population concentration, a process so evident in our own society. To build population in one place relies on the skill of storing foods. It was Lewis Binford in his study of the Alaskan Inuit who pointed out that storing the results of reindeer hunting made it possible to reduce the amount of travel in search of game and live in semi-permanent villages. Storage changes the archaeological evidence too. A reindeer killed in the summer might not be eaten until the next spring and in a different place. This would affect our interpretation of the seasonal information from animal bones. Equally importantly stores can be defended against other communities. And stored foods alter the needs placed on the labour force – rather than requiring a continuous supply of labour to find food, it becomes possible to concentrate the demand into a few weeks or months of the year.

Storage is often seen as an important feature of behaviour in 'complex' modern hunter-gatherers, but in the distant past it is difficult to demonstrate. We are used to the granaries and storage pits of farmers. Among hunters these are much rarer and usually pits are found only in the cold northern climates. The Kostenki sites on the Don River in Russia are good examples (see Figure 6.8). The alternative, which seems to have been the choice of Neanderthals, was to live at densities that made a secure living possible by always being among the herds of mammoth, rhino, bison, horse and reindeer. Modern humans played a different game. Where possible they settled down and based their security on their stores of food and their far-flung networks. The herds at some times of the year might be many miles away. The food-stores assured them of a supply, but if this failed their social networks allowed them to call on others.

We cannot be certain, but storing social relationships and organizing them according to the rules of kinship seem necessary developments for the remarkable geographical expansion of modern humans after

50,000–40,000 years ago. As important, we believe, as working out how to build a boat and paint a cave with animals.

Summary

A set of late revolutions in human evolution is not so much an impossible coincidence (as archaeologist Paul Mellars has termed it) as an impossible mystery. Our modern species *Homo sapiens* could not have acquired its large brain first, for nothing in particular, and then suddenly fired it into use 50,000–40,000 years ago. Evolution does not work that way. The large brain was there for something, if not the things that we do now. A huge benefit of the social brain ideas is that they explain a great deal that biology or archaeology on their own cannot. If we look at the last half-million years, we start from the puzzle that human beings had acquired bigger brains, and had become prominent on the landscape, but apparently – superficially – there was very little to foreshadow the huge cultural changes that came with anatomically modern humans within the last 100,000 years. The social brain tells us that social worlds and orders of intentionality had risen to new heights much further back – and then a close inspection does indeed begin to trace the archaeological will-o'-the-wisps that show these beginnings. The burials, the beads, the new technologies, the new social organization on the ground, all these speak of a deep transformation.

7

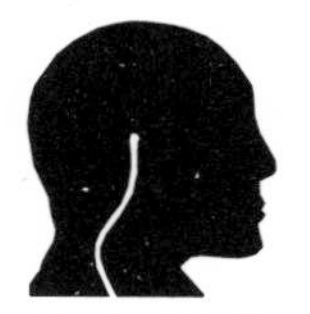

Living in big societies

Humans in the path of danger

Two milestones in human history were passed during the lifespan of the Lucy project. For the first time, in 2007, more people lived in cities than on the land, and in 2011, global population passed 7 billion. Archaeology puts these tipping points into perspective. At the end of the Ice Age 11,000 years ago one estimate for the size of the world's population is 7 million. There were no cities and people lived by fishing, gathering and hunting. They had art, ceremonies as shown by burials, and architecture that took the form of huts and villages. Their technologies included stone-tipped arrows, sickles and knives and included ground-stone bowls, mortars and pestles. A wide range of plant materials was also involved, woven into baskets and worn as clothes. The main solution to problems of fluctuating resources and hot tempers was to move, to literally walk away from the problem in the time-honoured pattern of fission and fusion still seen among today's hunters and gatherers.

There came a time when this would not work, because of a thousand-fold increase in population as a result of a revolution – the agricultural revolution. Archaeologists have long debated its origins, but how it came to be does not really affect our social brain arguments. The essence is that as food producers humans were able to sustain vastly greater populations. From about 11,000 years ago agricultural populations increased, and hunters and gatherers were on the wane.

When population was small and mobile the occurrence of natural disasters such as volcanic eruptions had little impact. This happened 12,900 years ago when one of the Rhineland volcanoes, the Laacher See, experienced a major eruption. Leading volcanologist Clive Oppenheimer describes the impact in *Eruptions that Shook the World* (2011). The plume peaked at heights of 35 km (22 miles) and the crater formed measures 2 km (over 1 mile) across. Ash fallout covered up to 300,000 sq. km (115,000 sq. miles) and due to the wind direction the plume stretched from central France to northern Italy and from southern Sweden to Poland. In the immediate vicinity of the crater are huge deposits of pumice that are mined today for building material. But, for all this devastation, there seems little evidence that human populations in the region suffered a major setback that led to desertion and extinction. They would have been able to move away from the area and, by calling on networks of social ties spread over the wider unaffected region, re-establish themselves. They resettled the worst-affected area within a few generations, which on an archaeological timescale looks instantaneous. The ability to move drew on the long-standing human traits of walking and having a social brain that deals with friends and strangers by establishing kin and sharing culture.

It was to be 'all change' for this state of affairs in the next 11,000 years. The option to move was increasingly restricted by the reliance on fields and flocks and the investment in towns, cities and the apparatus of political power. In January 2012 the *Daily Mail*, one of Britain's leading tabloid newspapers, ran a story about the Laacher See under the headline 'Is a supervolcano just 390 miles from London about to erupt?' Their scientific evidence was no more than an eruption is 'overdue', whatever that might mean. But suppose their scaremongering was right and that this time the wind was blowing from the east. Then the worst fears that tabloid Britain entertains about Europe would be realized: agriculture would be devastated, cities buried, transport brought to a halt (as air travel was in 2010 when a tiny ash cloud from Iceland covered the continent) and civil unrest guaranteed as people try to cope without smartphones.

A global population of 7 billion, concentrated in cities, courts disaster on a daily basis. Earthquakes and tsunamis have always been a feature of hominin evolution. But their potential impact has never been greater because of the numbers of people who now live in their path or beneath their smoking cones. If it really concerned us, we would relocate populations in danger to somewhere safer. But of course we can't, preferring to wait it out and deal with the consequences when they arise.

The home straight

What does this dance with danger tell us about one of the questions we posed in Chapter 1: will it ever be possible to say when hominin brains became human minds? This should be a question for philosophers, except they have little interest in the deep-time histories we have tackled in this book. What can we add to the picture?

We briefly explored different models of the mind in the box on pp. 106–07, contrasting the rational mind that solves problems with the relational one that works by building associations through the application of common social skills. Rationalism was one of the achievements of the European Enlightenment, while another was the rediscovery and excavation of Pompeii and Herculaneum. These sites show in graphic detail what a comparatively small eruption can do to two small towns. If the rational mind was a sign of the modern mind then surely, following these discoveries, Naples would have been downsized to a small fishing village to avert future disaster. Although we visit the sites and their museums and flock to exhibitions about their destruction, our rational minds do not lead us to the obvious conclusion – don't build here. Instead we are wedded to such dangerous places through family, social, economic and historical ties. Using our advanced theory of mind, many trust in religion to keep the firestorm at bay.

In our view the long trajectory, which eventually gives rise to the human mind that philosophers can pronounce on, principally involves the development of increasingly sophisticated social skills and a parallel increase in the size of social communities. As we have shown throughout this book, these skills are very ancient. They are not the sole prerogative of the 7 million *Homo sapiens* at the end of the Ice Age or the 7 billion living today. To a greater or lesser extent they are shared with earlier large-brained hominins such as *Homo heidelbergensis* and *Homo neanderthalensis*. They include the ability for mentalizing, to sympathize and empathize with others and to have an advanced theory of mind that allows access to another person's intentions; these are deeply buried in our ancestry and, as we have shown, can be traced back into our primate past. Several of them involve language and the use of objects in both symbolic and metaphorical ways. As a result there is no artifact or fossil ancestor that allows us to say categorically, 'This one had a modern mind, but that one didn't.' That is partly because we knew from the outset of the social brain project that coming up with a definition of a modern mind was a search for fool's gold. What we have done instead is present the mind as a set of social skills that throughout our evolutionary history was under constant selection. From our perspective these peaked with social communities of 150 and the cognitive abilities an individual required to manage such numbers. Dunbar's number of 150 has proved an immensely solid building block to create ever bigger and more elaborate buildings – which we have described with many examples using the term amplification. We have gone from 7 million to 7 billion in a few 'minutes' of our evolutionary history, and yet the core cognitive structure that regulates our social lives has remained the same, even though we have gone from the Stone Age to the Digital Age. What we have seen in this helter-skelter ride since the end of the Ice Age is the unleashing of human imagination to turn materials into novel forms of incredible variety and unprecedented quantities. But the minds behind this variety are essentially the same in terms of forging relations with others and solving the problems that all our ancestors have faced in making their way in the social world.

The social brain restated

The essence of our position is that for the last 11,000 years humans have lived in a brave new world of huge numbers, equipped only with the social skills and framework allowed by a far older biology. At its core the social brain is about group size. Social cognition is expensive. Where the ecology favours larger groups, there is a 'strain on the brain' – what we refer to as cognitive load. Over the course of 2 million years, evolution favoured ever larger brains in the genus *Homo*. The shift was a gradual response to evolutionary pressures. When humans then made sudden cultural and technological breakthroughs with enormous consequences, such as agriculture, the situation was different, and we could not respond to far larger societies with another spurt of brain growth.

Rather, life was the art of the possible. In the larger societies you could not know everyone. Previously groups would split when they became too large, but that option was now far harder. However, there would also be incentives for getting on with other groups. Ancient warfare is readily apparent, and in games played for high stakes, alliances between groups must have been crucial.

Much of the last 11,000 years has been about learning how to make the large numbers work, through using the old small numbers. Dunbar's number is still the constraint on the number of people that we can know and deal with. Most of the time we still live with small core groups of five, support groups of around 15, and band-size numbers of around 50. At the larger numbers, new responses were needed and we will look at three – religion, leadership and warfare. But first an archaeological surprise to give them some context.

The buried circles

Near the city of Urfa in southern Turkey is one of the most remarkable archaeological discoveries of the last 20 years. At Göbekli Tepe (the 'pot-bellied hill') archaeologist Klaus Schmidt and his international team have uncovered stone circles dating to 11,000 years ago (see Figure 7.1). These were built by people without domestic animals or crops and no pottery. Their living quarters have not yet been discovered. They were hunters and gatherers, yet they produced monumental architecture, giant T-shaped blocks up to 7 m (23 ft) tall – a sophisticated version of Stonehenge but 8000 years older – covered in carvings of animals including foxes, spiders and ducks. The pillars represent humans, as shown by the carved arms and hands on their sides. The multiple stone circles were then deliberately buried, creating the distinctive shape of the hill.

Göbekli Tepe challenges many archaeological stories. Two in particular come to mind in the context of the social brain. The first is the assumption that monumental architecture only appears with a settled life based on domestic animals and crops. And second, that to organize building works on this scale would need someone to lead the project: a man, inevitably, of vision and charisma not dissimilar to the popular image of an archaeologist running a major excavation in Greece or the Near East.

Linking up

Göbekli Tepe has forced archaeologists to think again about one of their favourite histories, the power of farming to change all aspects of society. But what are the wider implications of this remarkable site? What happened to our social brains with the appearance of farming? Did we evolve rapidly a new type of brain better able to process symbols that made living in villages and towns possible? This is the view of archaeologist Colin Renfrew who, in his book *Prehistory: Making of the Modern Mind* (2007), is impressed by the volume of new artifacts, artistic styles and architecture that arrives with the Neolithic revolution. These new things led, in his opinion, to a fundamentally different way of conceiving of other people and the world. For Renfrew it was a sedentary revolution, based on farming, which allowed people to engage with their material culture in ways not previously possible. This was the birth of the modern mind.

We are not so sure. The social brain of a hunter living 20,000 years ago in the wide-open spaces of the Near East does not seem to us to be necessarily any different from that of someone in the cramped quarters of Çatalhöyük, a Neolithic town of 8000 people in Turkey, first occupied in 7500 BC. Certainly there was a huge change in the types and variety of material things made by these hunters and farmers. But the stone circles at Göbekli Tepe, much older still and built by people without farming, challenges the traditional view of archaeologists that only by settling down and growing crops could advances in symbolic life be made.

When it comes to thinking big, the material worlds of those first farmers in the Near East certainly stand out. But we want to know if their social worlds were still underpinned by cognitive frameworks with a much older ancestry. Our analogy would be that the 21st-century mind is very different from the medieval mind; for example, science has replaced alchemy. But although the principles on which these two cultural worlds were based seem very different, we still expect that the cognitive building blocks by which people linked up in groups and communities remained similar.

Fiona Coward tested these assumptions during the Lucy project. She took the Near East as her region and examined the cultural inventories

Figure 7.1: *The amazing stone circles of Göbekli Tepe are 11,000 years old, predating the origins of agriculture and so provoking new thinking about the beginnings of complex societies.*

from no fewer than 591 archaeological sites dated to between 21,000 and 6000 BC. This huge sample spanned the change in climate from the cold of the Ice Age to the warmth of the Holocene. It also saw the transition from hunter to farmer. As a result population increased and people started to settle down. Was this the birth of the modern mind as Renfrew and others claim?

Coward's interest lay in examining social networks during this climatic and economic transition. She did this with a simple proposition: that networks can be traced through material remains. We have seen in earlier chapters how raw materials linked mobile peoples over large distances. Coward's database took this study of social ties using material culture to another level by cross-referencing many different types of artifacts. What this meant in practice was comparing the inventories of nearly 600 sites across the Near East in 15 time-slices of 1000 years each. Sites were socially tied together by the similarity of what they contained. Different varieties of material culture were tabulated: art, burial data, architecture and other built features, ground stone, hearths, chipped stone tools, ochre, ornaments and jewelry, shells and worked bone. These produced a matrix of relationships for each time-slice which measured the strength of the material ties between the people who lived at particular sites.

What did she find? After 13,000 BC, the network ties in the Near East started to centralize, with a small number of sites in the network becoming richer in material objects than others. This is to be expected as climates changed and the earliest examples of settling down are found. The networks of the first farmers were also more far flung than the earlier hunters, extending well beyond the distances the latter achieved with the movement of raw materials. The strength of ties between sites increased over time as a result of the sheer volume of new things that were now made on a regular basis. But interestingly this is not matched when the proportion of the links that could have been made is calculated. This proportion is described as the density of the network. Here the figure actually declined over time.

Archaeologists have always been impressed by the explosion in material culture during the Neolithic, so much so they called it a revolution. Some even see this as the moment that minds became modern. But they did this without thinking about the social networks which required such stuff and the cognitive abilities that underpinned its use. The explosion of material in the Neolithic, which we do not deny, does not however signal the arrival of the modern mind. Coward's study using a social brain perspective showed that an alternative was more likely, one that involved the amplification of an existing cognitive framework. Instead of a new, Neolithic, mind we find evidence for continuity of a common social brain.

The conclusion Coward draws is that material culture itself was the catalyst to thinking big. Hunters and farmers were after all the same human species: *Homo sapiens*. They may have seen the world differently, in the same way that the world-view of a medieval alchemist differed from that of a modern scientist. But when it came to building networks of contacts they used the same framework that had evolved to cope with the problems of too-little-time-in-the-day and the extra cognitive load if people are to be dealt with not as ciphers but in a socially meaningful way. The networks of artifacts, which so impress archaeologists comparing the Neolithic to the Palaeolithic, were a classic way of offloading the cost from the cognitive realm to the material world. The earliest hominins, with their smaller communities, simple stone tools, and not much else, started this path to complexity. What we see in the critical period from 21,000 to 6000 BC is an amplification of the materials that bind people together. By the end of this period, they were enmeshed in webs of culture, like Gulliver tied down by the ropes of the Lilliputians. The tiers of the social brain, such as Dunbar's number of 150, remained the same, but what was now possible, thanks to farming, was unimaginable growth in terms of the size of the population that now used this common cognitive framework for social interaction. Society had become truly complex, in ways that outdistanced those of the hunter and gatherer, especially in the numbers game, but social life was still based on some basic cognitive principles buried deep in our ancestry.

The power of charisma: religion, leadership and warfare

Were the monuments of Göbekli Tepe possible without a leader to organize the work? This would have been someone to direct the design, quarrying, carving and erection, not to mention the labour of back-filling the site. Just one of the circular enclosures needed 500 cu. m (17,500 cu. ft) of debris to cover it up. People had to buy into this grand design and the benefits for them are obscure to us, especially because the labour force lived by hunting and gathering. If they were farmers, archaeologists would be much less surprised by the monuments, which tells us a good deal about ourselves and the way we think about the past.

Does religion supply a motive and a possible explanation? Some of the very earliest evidence for more organized forms of religion comes from the Levantine settlements of the Neolithic period dating to around 8000 years ago, where some buildings have been interpreted as ritual sites – though later, more cautious archaeologists have shied away from over-interpreting the functions of these buildings. We are on decidedly firmer ground, however, by the subsequent Bronze Age (beginning around 5000 years ago), when the evidence for ritual sites and religious artifacts is considerably less controversial. We can speculate though, that since these can only exist once the conceptual apparatus needed to create them is in place, it is likely that the religions themselves as conceptual and ritual phenomena have much deeper roots.

Given the role of organized religion in coercing populations to toe the communal line, the most plausible explanation for the rise of these kinds of religion is the social and psychological stress imposed by large communities. If settled life is to survive, two problems need to be solved. The first is to dissipate the stresses of living in close proximity, where escape is difficult because the hunter's solution of walking away is far from easy. The second is to manage and control the activities of those looking for a 'free ride'. The activities of free-riders, getting something for nothing, threaten to undermine the communal contract on which large-scale village life necessarily depends.

Whether agriculture was a solution to an ecological problem or an accidental discovery that opened up novel opportunities is not an immediate issue for us. Our concern here is with the fact that living in settlements created new tensions that might have prevented the subsequent development of modern human life as we know it with all its cultural ramifications. Had humans not been able to find solutions to social constraints, the modern world would never have happened.

One feature that probably became of particular importance was the phenomenon of charismatic leaders, a topic that has been of particular interest to our collaborator Mark van Vugt (see his book, co-authored with Anjana Ahuja, *Selected: Why Some People Lead, Why Others Follow, and Why It Matters*, 2010). Although hunter-gatherer societies eschew any social distinctions (everyone is equal), charismatic leaders have come to play an especially important role in all post-Neolithic societies and in the doctrinal religions that these gave rise to. At one level, they provide leadership; at another, they provide a focus around which the community can gather.

At community sizes of 150, where everyone knows everyone else, leaders are probably unnecessary in the sense that, with only around 20 or 25 adult males in the group, arriving at a consensus on what to do is inevitably a great deal easier than in a community of 1500 individuals with ten times as many adult males. What exacerbates the situation in such large groups is the inevitable differentials among the males in terms of age, skills, experience and reputation – criteria on which males compete and which create rivalries. Too many males, and the risk of fractiousness and fragmentation increases because too many of them are of approximately equal standing. In such situations charismatic leaders provide a natural focus around which everyone can gather and sign up to whatever particular masthead the leader happens to espouse.

Charismatic leaders play a particularly important role in religion. Most of the doctrinal religions were founded by a single charismatic leader, from Zoroaster to Gautama Buddha and Jesus Christ to the Prophet Mohammed. Each of today's major religions began life as a minority sect of some predecessor religion. In fact, it is one of the features of religions that they seem to have a natural propensity to spawn sects. In most cases, these begin on a very small scale, but their survival ultimately depends on how quickly they can attract committed members. It is a feature of religious belief that it can rouse intense emotional commitment irrespective of the actual beliefs to which people sign up. It is as though the spirit world has a peculiar emotional hold over the human mind.

The skills that allow some individuals (usually men, but not always – think Joan of Arc) to become charismatic leaders continued to play an important role in the secular (i.e. political) world as the forces unleashed in the Neolithic revolution gradually gave rise to increasingly large political units over the ensuing millennia. By 3000 BC, city states had kings and viziers. Such people continue to contribute to the success of secular organizations of many different kinds, from businesses to schools to hospitals to charities.

Charismatic leaders become important in the context of super-large communities because they allow the group to impose some discipline on itself, to cut through the different levels of its huge numbers. Having a leader to coordinate and manage community behaviour has obvious advantages. But it comes at a cost since every individual has slightly different interests, and, in the end, an individual that emerges as a leader will always pursue the strategies that are more in line with his or her interests than anyone else's – not least because it is inevitable that everyone in the community will have slightly different views about the best thing to do. A charismatic leader who can persuade everyone else to act in concert, despite what each of them might prefer to do individually, might be beneficial in the long run if the net benefits of doing so exceed those that derive from everyone acting in their own personal interests.

Protection against marauders has been proposed as a major factor selecting for a steady increase in the size of polities (political units or societies), which is observable from the Neolithic onwards. Sociologist Alan Johnson and archaeologist Timothy Earle explicitly argued for this in their seminal book *The Evolution of Human Societies: From Foraging Group to Agrarian State* (1987). In their view, and that of an increasing number of evolutionary social scientists, warfare has been the principal driver of the ever larger settlements (villages, towns, cities, city states) that developed historically over the 7000–8000 years following the first villages. Raiding itself is a form of free-riding – gaining at someone else's expense. Why work, when for a very modest expenditure of energy and a small risk you can acquire all your needs from the products of others' toil? Defence against raiding becomes increasingly necessary as the advantages (and hence temptations) of this kind of free-riding increase. And organized military units with a clear line of command and organized direction are invariably more successful than ragtag armies that lack coordination. The charismatic leader comes into his or her own in just this kind of context.

In one sense, the rise of charismatic leaders takes us over the boundary of deep-history into history, and so beyond the main concerns of this book. Our point, however, is that the seeds of much that happened between the Neolithic and the modern day owe their origins to attempts to circumvent the stresses and strains created when humans first began to live in large settlements. Secular charismatic leaders, with their trappings of status such as bodyguards and conspicuous display, are similar to religious leaders with their priests, rites and ritual spaces. Both seek to coerce their followers to toe the communal line in order to make a more effective collaboration. Such collaborations are intrinsically unstable, because the communal line never suits everyone equally – some always imagine that they pay a disproportionate share of the common effort and eventually feel sufficiently aggrieved or empowered to revolt.

Just as religions constantly fragment into new sects built around charismatic leaders, so polities collapse in factional rivalry and revolution in the same way. As a species our solutions to the problems posed by the economic changes of the Neolithic are at best imperfect, mostly because the psychology we use to solve them is the psychology of mobile hunter-gatherers living in their dispersed communities, as our ancestors did for the previous 7 million years.

Picking and choosing

Powerful factors such as charismatic leadership and organized religion have played a large part in assisting the expansion of humanity. But there are other 'old' characteristics that also have helped to organize the new large societies. The numbers that we can deal with individually, as we have seen, do not change. 'Dunbar's number' is as much a reality for us as for the people of 100,000 years ago. What we have learnt to do is have selective personal networks. Two chimpanzees will have almost identical networks in their community – everyone knows everyone else. But in our world, we probably know most of the same people in the village or the street, and yet have completely different networks at work, in sports or in our hobbies. We switch selectively with ease.

The trades and hierarchies noted by Gordon Childe in the Neolithic hint at the origins of this switching. Merchants would want and need to deal with other merchants, perhaps hundreds of miles from home. Potters and metalworkers would want to discuss techniques with others who had the same skills. Leaders deal with leaders: when William the Conqueror divided England between 180 of his followers, he was following the principles of the social brain in brutal simplicity.

Charismatic leadership, religion and selective networks remain powerful forces whose deep-history we do not fully understand, as Göbekli Tepe shows. What we emphasize is that archaeology allows us to see the great sweep of human history often in close-up detail. Nobody who has seen an ancient city mound, a pyramid or the megalithic avenues of Carnac can fail to marvel at this tangible evidence of immense past social efforts. Whether collaborations of equals, or ventures collaboratively driven from the top, these collective monuments show how people regularly came together in numbers far exceeding their ancestors. These societies worked because their 'old' brains could deal with the new scale of social life. The selective hierarchical networks provided a framework for the growth of institutions and polities. At this point, the baton of documentation passes from archaeology to written history, sociology and psychology.

The technology of distributed minds: writing and texting

Fiona Coward, in her study of the networks of things in the Near East, makes the point that objects provided a framework for thinking big and then bigger. This is a good example of a distributed mind (see box on pp. 106–07) amplifying technology to meet new challenges, which in this case meant numbers of people. Before this explosion in things, we argued in Chapter 5 that language filled the grooming gap as brains and communities increased in size. Two further technologies have filled similar gaps, but this time to help organize and coordinate the huge populations of the last 11,000 years. These are writing and texting.

We would suggest that writing arose as a mechanism to ease the pressures of living in larger societies. Formal notation allowed commands to be sent, niceties of diplomacy to be shared and the accounts to be done. It was also the extension of communication that allowed formal institutions to come into being. In medieval times, St Francis of Assisi, an archetypal charismatic leader, cared nothing for property or organization. But the Catholic Church that he dealt with insisted on imposing some administrative control and the key to that was a written record.

Writing, indeed, has been the guarantor of historical continuity through the last 5000 years. We know the Chinese, Greeks and Romans far better than any other peoples of their age on account of their written records, which often detail their own struggles with organization and administration. The other great aids to social cohesion are improvements in communication and technology. The two come together in the archaeological evidence in a ship that sank at Sinan off the southern coast of Korea in AD 1323. On its way along the silk route from China to Japan, the ship was carrying thousands of ceramic pieces. The preserved goods, invoices and labels speak of merchants on their way from China to Japan, as well as monks making other journeys. Myriad other details are preserved for us throughout the last few thousand years, but few are as compelling as those that archaeology has recovered from time capsules such as Sinan, Pompeii and Herculaneum or Henry VIII's flagship the *Mary Rose*, and all give tantalizing glimpses into networks of the past.

Writing also made possible the Digital Age. And like writing, the arrival of the Internet has offered us the prospect of greatly widening our social circles. Indeed, this is exactly what the founders of many social networking sites initially promised. But has the promissory note on the tin can proved true? The answer seems to be a resounding 'no'. Despite the opportunity to create new connections at the click of a 'friending' button, in fact most people's Facebook pages list only between 100 and 250 names, as one study of 1 million Facebook pages recently revealed. In one of our studies, Tom Pollet and Sam Roberts asked whether regular Facebook users had larger social networks than casual Facebook users. They did not. Two other online studies that looked at message exchanges within email and Twitter communities also found that they too were typically between 100 and 200 people in size.

It seems that we cannot radically increase the number of relationships we have, even when technology would appear to allow it. The constraint on what we do is not time or our memory for the events in which we or our friends have been involved, but rather the limited space we have in our minds for friends. But there is another reason why the Internet won't allow us to have more, or more complex, social networks, and this is the fact that we don't find building new relationships via text-based media very satisfying. In a study carried out by Tatiana Vlahovic and Sam Roberts, subjects were asked to rate their satisfaction with interactions with their five best friends over a two-week period. Face-to-face interactions and those using Skype were rated as much more satisfying than interactions with the same friends via the phone, texting, instant messaging or social networking sites. In part, this was because these other media are slow and clunky – the reply to a witty comment, if it comes at all, follows an unavoidable delay. By the time it comes, the excitement of the exchange has faded. Skype seems to win out because it creates a sense of

being in the same room together, a sense of co-presence. And that means the pace of the exchange is much faster than it can ever be in text-based media: I see the smile already breaking on your face as I tell my joke. This is such a powerful effect that jokes we would find hilarious in the pub fall flat in an email. In addition, like face-to-face encounters, Skype is channel-rich (we hear *and* see), allowing communication redundancy.

What the digital world and the billions of people it now connects has done is amplify existing technologies of communication. One example will suffice: the number of words. In 1950 world population stood at 2.5 billion and the number of words in the English language numbered half a million. Some 50 years later, as Google NGRAM shows us, the number of words has grown to over 1 million, or 8000 new words a year. Over the same period population grew even faster to reach almost 7 billion. Word growth was powered by population growth and amplified by the possibilities of HTML, which was invented by Tim Berners-Lee in 1993. A combination of massive populations and a new technology cranked up, through the ancient process of amplification, the number of words we now use to connect us across our myriad social communities. The same has happened for images but at the moment, rather like archaeological objects, these cannot be so easily measured as words. One inevitable consequence of this will be the fragmentation of English into several new languages, because each person, and each community of interacting people, can cope with only a limited vocabulary (around 60,000 words) that they know and use regularly.

There is something else that is lacking in the digital world, and that is touch. Touch is a genuinely important part of our social world, even with strangers. All that fingertip grooming from our primate heritage is still very much with us. How we touch someone can say more about what we really mean than any words that we might utter. Words are slippery things: given the right emphasis, intonation or accompanying gesture, a sentence can mean exactly the opposite of what its words seem to say. So far, no one has cracked the problem of virtual touch, but if and when they do, it might represent a major advance in our ability to create super-large, well-bonded communities on the Internet.

Twitter has, of course, been a novel feature in the digital world and many have hailed it as a great democratizing innovation. Now we can create flash mobs and mass protests at a keystroke. In one sense, it is true: Twitter played a seminal role in coordinating the Arab Spring uprisings. But Twitter doesn't create relationships – it is more like a lighthouse flashing away in the dark, irrespective of whether there is a ship out there to see it. Those who have studied this phenomenon are clear that what lay at the root of the Arab Spring was not Twitter as such, but actual face-to-face networks between a small number of charismatic leaders who set the whole thing in motion. Twitter will allow us to coordinate a gathering at a particular place and time, but not a social or political movement. Like cultural icons, these arise from personal relationships between leaders and followers, and this has been the pattern throughout our history.

Epilogue: thinking big

With our backgrounds in evolutionary psychology and archaeology, our privilege, as authors of this book, is that we can stand back and take the long view. We can indulge in thinking big. We can see in the huge diversity of modern human behaviour ancient patterns that recur. The growth in size of the human brain through 2 million years is a response to pressures that were certainly social, and that operated within our species rather than in direct relation to the external world. This growth in brain size correlates with an increase in size of human groups – the observation that stands at the heart of the social brain idea, but far from the whole story. The correlation with brain size is bound to be approximate. The human physique is now less rugged than it was 30,000 years ago during the last Ice Age, and today we have slightly smaller brains. But all modern humans share a size and structure of brain that gives them common capabilities, and also makes them operate with common constraints. What really matter in our everyday lives are the things that have always mattered since the very dawn of time, the circumstances of our births and growing up, the friendships we make and the opportunities that we take.

We asked at the beginning of this book, 'When did hominin brains become human minds?' There is a catch in the question that we can trace back to Linnaeus' first labelling of our species as *Homo sapiens* – the 'clever humans'. *Homo* the genus appears more than 2 million years ago. We should probably refer to those ancestors as humans, rather than separating them from us with rather mystic names such as 'archaics'.

Figure 7.2: *The scale and complexity of modern life requires a new selectivity in the operation of the social brain: individuals use old network principles to chart their pathways through life.*

If our own species name *sapiens* is given on biological grounds, then genetics places it after the division from Neanderthal ancestors, and the skeletal evidence puts it somewhere in Africa about 200,000 years ago. If we ask whether we would invite the people of Herto (among the earliest anatomically modern humans) round to dinner, we have to admit how artificial such questions become. Even in the present day, modern human culture allows such diversity that the modern Western reader could not easily feel comfortable with the rigours of life among the Yanomamo of Brazil or even pastoralists on the Eurasian steppe. When the people from small-scale societies are plucked out into larger ones, they too feel confusion – as did the native British prince Caractacus, who, dragged in chains to Rome in the 1st century AD, declared to his captors, 'You who have this [Rome with its grand buildings]... why do you come to our poor hovels?' These differences do not undercut the fundamental unity of cultural capability in all living humans. If we return to the Herto question, they too must be with us. The diaspora of modern humans is traceable back to common roots less than 200,000 years ago, and in a sense we are all cousins, if very distant cousins. The 200,000-year expanding cone represents humanity in the fullest sense.

Can the social brain help resolve the next big paradox – that these modern humans did relatively little that was modern for their first

100,000 years? Even were we to subscribe fully to a modern 'human revolution' 50,000 years ago – which in this book we have not done – there is a further 40,000 years and more before humans scale up to farming, villages, cities and civilization. The social brain helps us to see that this scaling up was only possible because humans had already acquired the orders of intentionality, the language, the means of bonding, that would allow them to live in larger societies. Farming, however it began, would inevitably generate the larger numbers. From then on, operation of the organizational principles outlined in these pages became inevitable in the species that we call *Homo sapiens*.

A dramatic contemporary example of the forces for change is the Internet. Only a very small number of people foresaw its potential power (and no-one could have foreseen all its consequences). At one level, the Internet represents a human behaviour-changing phenomenon on a world- and even species-wide scale, all delivered in one generation since 1993. As individual citizens, we are free to ignore this electronic revolution, but that is becoming less and less possible day by day. It pervades our lives, from the home to the workplace, from the beach to the restaurant.

At another level, however, the social networks we have been discussing in this book still seem to rule. Internet giants such as Facebook and Twitter depend for their success on the human desire for social contact and networking. And in the political sphere, those who represent governments in a sense 'personify' them – making them personal. Meetings such as the G8 or G20 are the tip of this iceberg. It is interesting to note that there are around 200 countries in the world (within the natural range of variation around Dunbar's number), but running down the scale, the top 12 of them, and to a lesser extent the top 36, have by far the biggest say. They are networked by their heads of state, but also by conclaves of ministers and officials who have their own select networks. The world functions because of this networking. Part of what we can show by studying the deep-history of the social brain is that networking exists necessarily on certain scales and with certain intensities. It works now only because of the past, shaped by the fireside, in the hunt and on the grasslands where we evolved. Social networking captures the techno-chic of the moment, that 'Vorsprung durch Technik'. But for all its glitz as the latest new technology to bind, bond and communicate, the principles it enshrines originated deep in our evolutionary past.

Selected reading

Chapter 1 – Psychology meets archaeology

For a more detailed introduction to the ideas behind the Lucy project see Dunbar & Shultz (2007) and the edited volumes by project members Dunbar, Gamble & Gowlett (2010), Allen et al. (2008) and Dunbar, Gamble & Gowlett (2013).

Two shorter papers by the authors of this book set out the need for a social approach and the framework that we elaborate on throughout *Thinking Big*. These are Gamble, Gowlett & Dunbar (2011) and Gowlett, Gamble & Dunbar (2012). The idea of social intelligence applied to monkeys and apes, which underpins so much of the social brain, is well covered in the selection of papers in Byrne & Whiten (1988). The evidence for the social brain hypothesis itself is given in Dunbar (1992).

The book from which the project took its name, *From Lucy to language* by Johanson & Edgar (1996), remains one of the best-illustrated accounts of hominin evolution, although there have been a number of major fossil finds since it was first published almost 20 years ago. Stringer & Andrews (2011) brings the story up to date and is particularly strong on the origins of *Homo sapiens*, so-called modern humans. The breaking of the deep-time barrier, and the re-finding of the handaxe that made this possible, is told by Gamble & Kruszynski (2009).

Allen, N.J., Callan, H., Dunbar, R.I.M., & James, W. (eds). 2008. *Early human kinship: from sex to social reproduction*. Oxford: Blackwell.

Byrne, R.W., & Whiten, R. 1988. *Machiavellian intelligence: social expertise and the evolution of intellect in monkeys, apes and humans*. Oxford: Clarendon Press.

Dunbar, R.I.M. 1992. 'Neocortex size as a constraint on group size in primates.' *Journal of Human Evolution* 22: 469–93.

Dunbar, R.I.M., Gamble, C., & Gowlett, J.A.J. (eds). 2010. *Social brain and distributed mind*. Oxford: Oxford University Press. Proceedings of the British Academy 158.

Dunbar, R.I.M., Gamble, C., & Gowlett, J.A.J. (eds). 2013. *The Lucy project: benchmark papers*. Oxford: Oxford University Press.

Dunbar, R.I.M. & Shultz, S. (eds). 2007. 'Evolution in the social brain.' *Science* 317: 1344–47.

Gamble, C., Gowlett, J.A.J. & Dunbar, R.I.M. 2011. 'The social brain and the shape of the Palaeolithic.' *Cambridge Archaeological Journal* 21: 115–35.

Gamble, C., & Kruszynski, R. 2009. 'John Evans, Joseph Prestwich and the stone that shattered the time barrier.' *Antiquity* 83: 461–75.

Gowlett, J.A.J., Gamble, C., & Dunbar, R.I.M. 2012. 'Human evolution and the archaeology of the social brain.' *Current Anthropology* 53: 693–722.

Johanson, D.C., & Edgar, B. 1996. *From Lucy to language*. New York: Simon & Schuster.

Stringer, C., & Andrews, P. 2011 (2nd edition). *The complete world of human evolution*. London: Thames & Hudson.

Chapter 2 – What it means to be social

A general overview of the structure and function of human societies can be found in Dunbar (1996, 2008). The original proposal for Dunbar's number can be found in Dunbar (1993), and the evidence for it is summarized there and in Dunbar (2008). Evidence for the layered structure of social networks and the 'rule of 3' is given by Zhou et al. (2005), and the fact that this extends to the structure of human organizations, including the army, is summarized in Dunbar (2008, 2011, 2012) and Lehmann, Lee & Dunbar (2013). For information relating to the functions of the different grouping layers, the role of social interaction in maintaining them, and the prosocial consequences, see Roberts & Dunbar (2011), Sutcliffe et al. (2012) and Curry, Roberts & Dunbar (2013). The relationship between network size, emotional closeness and brain volumes is discussed in Stiller & Dunbar (2007), Dunbar (2008) and Powell et al. (2012).

Curry, O., Roberts, S.B.G., & Dunbar, R.I.M. 2013. 'Altruism in social networks: evidence for a "kinship premium".' *British Journal of Psychology* 104: 283–95.

Dunbar, R.I.M. 1993. 'Coevolution of neocortex size, group size and language in humans.' *Behavioral and Brain Sciences* 16: 681–735.

Dunbar, R.I.M. 1996. *Grooming, gossip and the evolution of language*. London: Faber & Faber.

Dunbar, R.I.M. 2004. *The human story*. London: Faber & Faber.

Dunbar, R.I.M. 2008. 'Mind the gap: or why humans aren't just great apes.' *Proceedings of the British Academy* 154: 403–23.

Dunbar, R.I.M. 2011. 'Constraints on the evolution of social institutions and their implications for information flow.' *Journal of Institutional Economics* 7: 345–71.

Dunbar, R.I.M. 2012. 'Instant expert #21: evolution of social networks.' *New Scientist* 214: 1–8.

Dunbar, R.I.M., Baron, R., Frangou, A., Pearce, E., van Leeuwen, E.J.C., Stow, J., Partridge, P., MacDonald, I., Barra, V., & van Vugt, M. 2012. 'Social laughter is correlated with an elevated pain threshold.' *Proceedings of the Royal Society, London* 279B: 1161–67.

Dunbar, R.I.M., Kaskatis, K., MacDonald, I., & Barra, V. 2012. 'Performance of music elevates pain threshold and positive affect.' *Evolutionary Psychology* 10: 688–702.

Lehmann, J., Lee, P.C., & Dunbar, R.I.M. 2013. 'Unravelling the evolutionary function of communities.' In Dunbar, R.I.M, Gamble, C., & Gowlett, J.A.J. (eds), *The Lucy project: benchmark papers*, Oxford: Oxford University Press, 245–76.

Powell, J., Lewis, P.A., Roberts, N., García-Fiñana, M., & Dunbar, R.I.M. 2012. 'Orbital prefrontal cortex volume predicts social network size: an imaging study of individual differences in humans.' *Proceedings of the Royal Society, London* 279B: 2157–62.

Roberts, S.B.G., & Dunbar, R.I.M. 2011. 'The costs of family and friends: an 18-month longitudinal study of relationship maintenance and decay.' *Evolution and Human Behavior* 32: 186–97.

Stiller, J., & Dunbar, R.I.M. 2007. 'Perspective-taking and memory capacity predict social network size.' *Social Networks* 29: 93–104.

Sutcliffe, A.J., Dunbar, R.I.M., Binder, J., & Arrow, H. 2012. 'Relationships and the social brain: integrating psychological and evolutionary perspectives.' *British Journal of Psychology* 103: 149–68.

Zhou, W-X., Sornette, D., Hill, R.A., & Dunbar, R.I.M. 2005. 'Discrete hierarchical organization of social group sizes.' *Proceedings of the Royal Society, London* 272B: 439–44.

Chapter 3 – Ancient social lives

In this chapter we reach out to the context of surrounding disciplines. In the shaping of archaeology, Childe and his books feature in many accounts, including Daniel & Renfrew (1988). The importance of hunter-gatherers was reasserted in the famous *Man the hunter* (Lee & DeVore 1968), and both peoples and anthropologists are treated in depth in Lee & Daly (1999). A compendium of information on hunter-gatherers by Binford (2001) is also indispensable. An evolutionary view of the role of social anthropology is provided by Barnard (2011). We list below other important contributions to anthropology, to other broad aspects of human behaviour, and also to primatology, which has altered many perceptions about human origins, most particularly through the study of living apes.

Barnard, A. 2011. *Social anthropology and human origins*. Cambridge: Cambridge University Press.

Binford, L.R. 2001. *Constructing frames of reference: an analytical method for archaeological theory building using ethnographic and environmental datasets*. Berkeley: University of California Press.

Boesch, C. 2012: *Wild cultures: a comparison between chimpanzee and human cultures*. Cambridge: Cambridge University Press.

Childe, V.G. 1951. *Social evolution*. London: Watts.

Daniel, G., & Renfrew, C. 1988. *The idea of prehistory*. Edinburgh: Edinburgh University Press.

d'Errico, F., & Backwell, L. (eds). 2005. *From tools to symbols: from early hominids to modern humans*. Johannesburg: Witwatersrand University Press.

Dillingham, B., & Carneiro, R.L. (eds). 1987. *Leslie A. White's ethnological essays*. Albuquerque: University of New Mexico Press.

Gowlett, J.A.J. 2009. 'The longest transition or multiple revolutions? Curves and steps in the record of human origins.' In Camps, M., &

Chauhan, P. (eds), *A sourcebook of Palaeolithic transitions: methods, theories and interpretations*, Berlin: Springer Verlag, 65–78.
Gowlett, J.A.J. 2009. 'Artefacts of apes, humans and others: towards comparative assessment.' *Journal of Human Evolution* 57: 401–10.
Lee, R.B., & Daly, R. 1999. *The Cambridge Encyclopaedia of Hunters and Gatherers*. Cambridge: Cambridge University Press.
Lee, R.B., & Devore, I. (eds). 1968. *Man the hunter*. Chicago: Aldine.
McGrew, W.C., Marchant, L.F., & Nishida, T. (eds). 1996. *Great ape societies*. Cambridge: Cambridge University Press.
Moss, C. 1988. *Elephant memories*. London: Montana.
Nelson, E., Rolian, C., Cashmore, L., & Shultz, S. 2012. 'Digit ratios predict polygyny in early apes, *Ardipithecus*, Neanderthals and early modern humans but not in *Australopithecus*.' *Proceedings of the Royal Society, London* 278B: 1556–63.
Nelson, E., & Shultz, S. 2010. 'Finger length ratios (2D:4D) in anthropoids implicate reduced prenatal androgens in social bonding.' *American Journal of Physical Anthropology* 141: 395–405.
Roberts, S.G.B. 2010. 'Constraints on social networks.' In Dunbar, R.I.M., Gamble, C., & Gowlett, J.A.J. (eds), *Social brain and distributed mind*, Oxford: Oxford University Press, Proceedings of the British Academy 158, 115–34.
Runciman, W.G. 2009. *The theory of cultural and natural selection*. Cambridge: Cambridge University Press.
Sahlins, M. 1974. *Stone Age economics*. London: Tavistock Publications.
Shennan, S.J. 2002. *Genes, memes and human history: Darwinian archaeology and cultural evolution*. London: Thames & Hudson.

Chapter 4 – Ancestors with small brains

In this chapter we turn to the evidence of the earliest hominins and their descendants the australopithecines. Most new hominin remains are published in the major journals *Nature* and *Science*, with greater detail given in the *Journal of Human Evolution*, *The American Journal of Physical Anthropology* and the French journal *Palevol*. The East African projects are also published in large volumes, such as the Olduvai series (e.g. Leakey 1971). Reynolds & Gallagher (2012) includes many papers giving details of the African record. Early Asia is treated in Dennell (2009). The oldest finds from Dmanisi are given in Lordkipanidze et al. (2013).

The brain sizes of fossil hominins, as well as their classifications, can be found in Aiello & Dunbar (1993), where the methods for calculating group size are also explained. Updated calculations with revised equations are given in Dunbar (2009). Wood & Lonergan provide a clear overview of what fossil ancestors have been called and how to group them by lumping and splitting. A classic paper by Lovejoy (1981) marks the modern way of looking at fossil and archaeological remains not just as something to describe but instead as hominins with cultural and physical adaptations that evolved to meet a variety of selection pressures. The recent descriptions of the important fossil *Ardipithecus* in a special issue of *Science* (vol. 326, 2009) show the value of this approach. A comprehensive overview of the earliest stone tools is provided by Schick & Toth (1993).

Aiello, L., & Dunbar, R.I.M. 1993. 'Neocortex size, group size and the evolution of language.' *Current Anthropology* 34: 184–93.

Brain, C.K. 1981. *The hunters or the hunted? An introduction to African cave taphonomy*. Chicago: University of Chicago Press.

de la Torre, I. 2011. 'The origins of stone tool technology in Africa: a historical perspective.' *Philosophical Transactions of the Royal Society of London B* 366: 1028–37.

Dennell, R. 2009. *The Palaeolithic settlement of Asia*. Cambridge: Cambridge University Press.

de Waal, F. 1982. *Chimpanzee politics*. London: Johnathon Cape.

de Waal, F. 2006. *Primates and philosophers: how morality evolved*. Princeton: Princeton University Press.

Domínguez-Rodrigo, M. (ed.). 2012. *Stone tools and fossil bones: debates in the archaeology of human origins*. Cambridge: Cambridge University Press.

Dunbar, R.I.M. 2009. 'Why only humans have language.' In Botha, R., & Knight, C. (eds), *The prehistory of language*, Oxford: Oxford University Press, 12–35.

Gamble, C. 2014. *Settling the Earth: the archaeology of deep human history*. Cambridge: Cambridge University Press.

Holdaway, S., & Stern, N. 2004. *A record in stone: the study of Australia's flaked stone artefacts*. Melbourne: Victoria Museum.

Leakey, M.D. 1971. *Olduvai Gorge: excavations in Beds I and II 1960–1963*. Cambridge: Cambridge University Press.

Lordkipanidze, D., Ponce de Leon, M.S., Margvelashvili, A., Rak, Y., Rightmire, G.P., Vekua, A., & Zollikofer, C.P.E. 2013. 'A complete skull from Dmanisi, Georgia, and the evolutionary biology of Early Homo.' *Science* 342: 326–31.

Lovejoy, C.O. 1981. 'The origin of man.' *Science* 211: 341–50.

Lycett, S.J., & Chauhan, P.R. 2010. *New perspectives on old stones: analytical approaches to Paleolithic technologies*. Dordrecht: Springer.

Reynolds, S.C., & Gallagher, A. (eds). 2012. *African genesis: perspectives on hominin evolution*. Cambridge: Cambridge University Press.

Schick, K.D., & Toth, N. 1993. *Making silent stones speak: human evolution and the dawn of technology*. New York: Simon & Schuster.

Suwa, G., Asfaw, B., Kono, R.T., Kubo, D., Lovejoy, C.O., & White, T.D. 2009. 'The *Ardipithecus ramidus* skull and its implications for hominid origins.' *Science* 326: 68e1–7.

Wood, B., & Lonergan, N. 2008. 'The hominin fossil record: taxa, grades and clades.' *Journal of Anatomy* 212: 354–76.

Chapter 5 – Building the human niche: three crucial skills

In this chapter we concentrated on three main themes: handaxes, fire and language. Goren-Inbar & Sharon (2006) does justice to the first, while Machin's paper of 2009 gives a view of their social implications. Fire is treated in several volumes and papers in the list below. Deacon (1997) gives an excellent overview of the context of language evolution, also writing in our conference volume (Dunbar et al. 2010). Aiello & Dunbar (1993) provide the classic foundation of relationships between brain size, group size and the origins of language. MacLarnon & Hewitt (2004) and Martinez, Rosa & Arsuaga (2004) give insights into anatomical evidence that may go with the emergence of language.

Aiello, L., & Dunbar, R.I.M. 1993. 'Neocortex size, group size and the evolution of language.' *Current Anthropology* 34: 184–93.

Aiello, L., & Wheeler, P. 1995. 'The expensive-tissue hypothesis: the brain and the digestive system in human and primate evolution.' *Current Anthropology* 36: 199–221.

Alperson-Afil, N., & Goren-Inbar, N. 2010. *The Acheulian site of Gesher Benot Ya'aqov. Volume II: Ancient flames and controlled use of fire*. Dordrecht: Springer.

Deacon, T. 1997. *The symbolic species: the co-evolution of language and the human brain*. London: Allen Lane, The Penguin Press.

Dunbar, R.I.M., Gamble, C., & Gowlett, J.A.J. (eds). 2010. *Social brain and distributed mind*. Oxford: Oxford University Press. Proceedings of the British Academy 158.

Goren-Inbar, N., & Sharon, G. (eds). 2006. *Axe age: Acheulian tool-making from quarry to discard*. London: Equinox.

Gowlett, J.A.J. 2010. 'Firing up the social social brain.' In Dunbar, Gamble & Gowlett, 341–66.

Gowlett, J.A.J. 2011. 'The empire of the Acheulean strikes back.' In Sept & Pilbeam, 93–114.

Gowlett, J.A.J., & Wrangham, R.W. 2013. 'Earliest fire in Africa: the convergence of archaeological evidence and the cooking hypothesis.' *Azania: Archaeological Research in Africa* 48: 5–30.

Isaac, B. (ed). 1989. *The archaeology of human origins: papers by Glynn Isaac*. Cambridge: Cambridge University Press.

Lycett, S.J., & Gowlett, J.A.J. 2008. 'On questions surrounding the Acheulean "tradition".' *World Archaeology* 40(3): 295–315.

Lycett, S.J., & Norton, C.J. 2010. 'A demographic model for Palaeolithic technological evolution: the case of East Asia and the Movius Line.' *Quaternary International* 211: 55–65.

Machin, A. 2009. 'The role of the individual agent in Acheulean biface variability: a multi-factorial model.' *Journal of Social Archaeology* 9(1): 35–58.

MacLarnon, A.M., & Hewitt, G.P. 2004. 'Increased breathing control: another factor in the evolution of human language.' *Evolutionary Anthropology* 13: 181–97.

Martinez, M., Rosa, M., Arsuaga, J.-L., et al. 2004. 'Auditory capacities in Middle Pleistocene humans from the Sierra de Atapuerca in Spain.' *Proceedings of the National Academy of Sciences USA* 101: 9976–81.

Sept, J., & Pilbeam, D. (eds). 2011. *Casting the net wide. Studies in honor of Glynn Isaac and his approach to human origins research.*

American School of Prehistoric Research Monograph Series. Harvard: Peabody Museum.
Walker, A., & Leakey, R. (eds). 1993. *The Nariokotome* Homo erectus *skeleton*. Cambridge, MA: Harvard University Press.
Wrangham, R.W. 2009. *Catching fire: how cooking made us human*. London: Profile Books.

Chapter 6 – Ancestors with large brains

The dramatic expansion in technology and culture that took place 50,000 years ago is often described as a human revolution. However, in this chapter we trace much earlier roots for these distinctive facets of humanity using a social brain perspective. For music and kinship we recommend two edited works by Bannan (2012) and Allen et al. (2008). Lewis-Williams (2003) remains an inspiration for a cognitive approach to the past by examining the link between the oldest art and shamanism, while the role of emotions in human evolution is explored by Turner (2000). The impact of these evolutionary changes on our distribution as a global species is presented by Gamble (2014). An invaluable guide to the archaeogenetics of past human populations can be found in Oppenheimer (2004).

Allen, N.J., Callan, H., Dunbar, R.I.M., & James, W. (eds). 2008. *Early human kinship: from sex to social reproduction*. Oxford: Blackwell.
Bannan, N. (ed.). 2012. *Music, language, and human evolution*. Oxford: Oxford University Press.
Evans, D. 2004. 'The search hypothesis of emotion.' In Evans & Cruse, 179–91.
Evans, D., & Cruse, P. (eds). 2004. *Emotion, evolution and rationality*. Oxford: Oxford University Press.
Gamble, C. 2007. *Origins and revolutions: human identity in earliest prehistory*. New York: Cambridge University Press.
Gamble, C. 2012. 'When the words dry up: music and material metaphors half a million years ago.' In Bannan, 85–106.
Gamble, C. 2014. *Settling the Earth: the archaeology of deep human history*. Cambridge: Cambridge University Press.
Green, R.E., Krause, J., Briggs, A.W., et al. 2010. 'A draft sequence of the Neandertal genome.' *Science* 328: 710–22.

Lewis-Williams, D. 2003. *The mind in the cave*. London: Thames & Hudson.

Mellars, P., Bar-Yosef, O., Stringer, C., & Boyle, K.V. (eds). 2007. *Rethinking the human revolution*. Cambridge: McDonald Institute.

Mithen, S. 2005. *The singing Neanderthals: the origins of music, language, mind and body*. London: Weidenfeld & Nicolson.

Noonan, J.P., Coop, G., Kudaravalli, S., et al. 2006. 'Sequencing and analysis of Neanderthal genomic DNA.' *Science* 314: 1113–18.

Oppenheimer, S. 2004. *Out of Eden: the peopling of the world*. London: Robinson.

Papagianni, D., & Morse, M.A. 2013. *The Neanderthals rediscovered: how modern science is rewriting the story*. London: Thames & Hudson.

Pearce, E., Stringer, C., & Dunbar, R.I.M. 2013. 'New insights into differences in brain organization between Neanderthals and anatomically modern humans.' *Proceedings of the Royal Society, London* 280B: doi: 10.1098/rspb.2013.0168.

Pettitt, P.B. 2011. *The Palaeolithic origins of human burial*. London: Routledge.

Rightmire, G.P. 2004. 'Brain size and encephalization in early to mid-Pleistocene *Homo*.' *American Journal of Physical Anthropology* 124: 109–23.

Shryock, A., Trautmann, T., & Gamble, C. 2011. 'Imagining the human in deep time.' In Shryock, A., & Smail, D.L. (eds), *Deep history: the architecture of past and present*, Berkeley: University of California Press, 21–52.

Stringer, C. 2006. *Homo britannicus: the incredible story of human life in Britain*. London: Penguin.

Stringer, C., & Gamble, C. 1993. *In search of the Neanderthals: solving the puzzle of human origins*. London: Thames & Hudson.

Turner, J.H. 2000. *On the origins of human emotions: a sociological inquiry into the evolution of human affect*. Stanford: Stanford University Press.

Wood, B., & Lonergan, N. 2008. 'The hominin fossil record: taxa, grades and clades.' *Journal of Anatomy* 212: 354–76.

Climate change, increased natural hazards and the explosion of global population are well covered from the perspectives of archaeology and vulcanology by Fagan (2008) and Oppenheimer (2011). The connectedness of worlds, ancient and modern, as well as the power of small things to tell big histories, is illustrated by MacGregor (2010).

Boyer, P. 2001. *Religion explained: the human instincts that fashion gods, spirits and ancestors*. London: William Heinemann.

Coward, F. 2010. 'Small worlds, material culture and ancient near eastern social networks.' In Dunbar, R.I.M., Gamble, C., & Gowlett, J.A.J. (eds), *Social brain and distributed mind*, Oxford: Oxford University Press, Proceedings of the British Academy 158, 449–80.

Coward, F., & Gamble, C. 2008. 'Big brains, small worlds: material culture and the evolution of mind.' *Philosophical Transactions of the Royal Society of London B* 363: 1969–79.

Dietrich, O., Heun, M., Notroff, J., Schmidt, K., & Zarnkow, M. 2012. 'The role of cult and feasting in the emergence of Neolithic communities. New evidence from Göbekli Tepe, south-eastern Turkey.' *Antiquity* 86: 674–95.

Fagan, B.M. 2008. *The great warming: climate change and the rise and fall of civilizations*. London: Bloomsbury Press

Foley, R., & Gamble, C. 2009. 'The ecology of social transitions in human evolution.' *Philosophical Transactions of the Royal Society of London B* 364: 3267–79.

MacGregor, N. 2010. *The history of the world in 100 objects*. London: British Museum Press.

Oppenheimer, C. 2011. *Eruptions that shook the world*. Cambridge: Cambridge University Press.

Trigger, B.G. 2003. *Understanding early civilizations: a comparative study*. Cambridge: Cambridge University Press.

Van Vugt, M., & Ahuja, A. 2011. *Selected: why some people lead, why others follow, and why it matters*. Toronto: Random House.

Sources of illustrations

Full bibliographic references not given below may be found in the Selected Reading on pp. 205–15. Silhouettes drawn by John Gowlett.

1.1 After Gamble 2007: fig. 8.1; **1.2** After Roberts 2008: fig. 1; **1.3** Robin Dunbar; **1.4** Bibliothèques d'Amiens Métropole; **1.5** Natural History Museum/Science Photo Library; **1.6** Robin Dunbar; **1.7** Courtesy Jeanne Sept; **1.8** © Staffan Widstrand/Corbis; **1.9** John Gowlett; **2.1** © Marius Coetzee; **2.2** Robin Dunbar; **2.3** Robin Dunbar; **2.4** Robin Dunbar; **3.1** (above) Nelson et al. 2011: fig. 2 © 2011, The Royal Society; **3.1** (below) Anna Rassadnikova/iStockphoto.com; **3.2** After Gamble, Gowlett & Dunbar 2011: fig. 1; **3.3** John Gowlett; **3.4** John Gowlett after Gowlett in Camps & Chauhan 2009: fig. 1; **3.5** After Gamble, Gowlett & Dunbar 2011: fig. 2; **3.6** AP/Press Association Images; **3.7** John Rendall/AP/Press Association Images; **4.1** Foley & Gamble 2009: fig. 2 © 2009, The Royal Society; **4.2** © J. Paul Getty Trust and the Tanzanian Department of Antiquities; **4.3** John Reader/Science Photo Library; **4.4** (left to right) John Reader/Science Photo Library; © John Sibbick; **4.5** © Kathelijne Koops; **4.6** John Gowlett; **4.7** John Gowlett; **4.8** © Peter Pfarr, Lower Saxony State Service for Cultural Heritage; **5.1** After Gowlett, Gamble & Dunbar 2012: fig. 4; **5.2** John Gowlett; **5.3** John Gowlett; **5.4** John Gowlett; **5.5** John Gowlett; **5.6** John Gowlett; **Table 5.1** Data after Peters, C., & O'Brien, E., 1981. 'The early hominid plant-food niche.' *Current Anthropology* 22, 127–40.; **5.7** John Gowlett; **5.8** John Gowlett after Gowlett 2010: fig. 17.4; **5.9** © John Sibbick; **5.10** John Gowlett after Gowlett 2010: fig. 17.2; **Table 6.1** Data from Rightmire 2004; Wood & Lonergan 2008; **6.1** After Noonan et al. 2006: fig. 6; **6.2** David Lewis-Williams/Rock Art Research Institute, University of Witwatersrand, Johannesburg; **6.3** Timeline after McBrearty, S., & Stringer, C., 2007. 'The coast in colour.' *Nature* 449: 793–94: fig. 2. Photos (top to bottom) Courtesy Peter Mitchell; David Roberts, Bangor University; Courtesy Christopher Henshilwood; Courtesy Christopher Henshilwood and Francesco d'Errico; **6.4** Courtesy Christopher Henshilwood and Francesco d'Errico; **6.5** Courtesy Christopher Henshilwood; **6.6** Drawing Rob Read; **6.7** John Reader/Science Photo Library; **6.8** Ira Block/National Geographic Creative; **6.9** After M. Boule, *L'Homme fossile de la Chapelle-suz-Saints*, Annales de Paléontologie, 1911–13; **7.1** © N. Becker/DAI; **7.2** James R. Gowlett.

Index

Page numbers in *italic* refer to illustrations and tables.